Simple Plant Propagation

OTHER BARRON'S GARDENING BOOKS

Simple Plant Propagation

R.C.M. Wright

BARRON'S/Woodbury New York

First published in Great Britain
1974 by Ward Lock Limited, 116
Baker Street, London, W1M 2BB,
a member of the Pentos Group

Designed by John Munday

International Standard Book Number 0–8120–0795–6
Library of Congress Catalog Card Number 77–70396

American Edition Published in 1977 by
Barron's Educational Series, Inc.
113 Crossways Park Drive
Woodbury, New York 11797

Printed and bound in Great Britain by
Cox and Wyman Limited, London,
Fakenham and Reading

Contents

Preface

Even the casual observer cannot fail to be impressed by the amazing diversity of plant life. This variation has largely been brought about by the ceaseless struggle for existence that occurs everywhere in nature. Therefore many of the modifications or adaptions have considerable bearing on a plant's perpetuation and increase.

Familiar examples of this are the runners of the strawberry, the tubers of the potato and the rhizomes of the iris–all of which are modified stems adapted to serve as a natural means of reproduction.

Adaptations in plants often relate to seed propagation which is a common natural method. The bright colours of many flowers and the fragrance of others are designed to attract insects, so that they may carry out the vital task of pollination.

Plant cultivators throughout the ages have learned a great deal about propagation by observing nature's methods and by trying to imitate them.

1
How plants increase

There are two distinct methods of plant increase: a) **seed** (sexual) and b) **vegetative** (asexual).

Seeds often afford a cheap and convenient method of raising large numbers of plants. Such plants are usually healthy, for the risk of transmitting disease from parents to offspring is much less with seed than when vegetative methods are employed. A definite disadvantage with seed is slowness to reach maturity and seedlings may need nursing for several years before they 'pay their keep' in the garden. This applies for example in the case of bulbs and fruit trees. Obviously in such circumstances vegetative methods of increase should be used if possible.

Vegetative methods simply involve isolating portions of a plant–whether of the stem, root or leaves–and inducing them to grow and develop into separate individuals. The resulting plants are truly 'chips off the old block' and not distinct original specimens as are seedlings. Because the function of sex is not involved, vegetative propagation is termed asexual.

The new plants are part of the old and this ensures absolute uniformity and conformity of type. This is the principal advantage of the method. Another advantage is that many plants so raised develop rapidly into mature and productive plants which do not require careful nursing in the early stages so essential with seedlings, and can compete on better terms with weeds and plant pests.

It is helpful to note which type of propagation predominates in the various groups of garden plants. Annual and biennial flowers are, of course, normally increased by seed.

Natural asexual propagation in the common garden buttercup. This normally increases by runners.

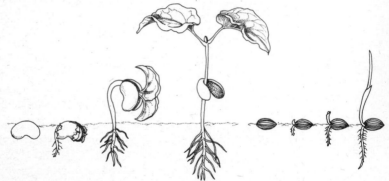

The germination of seeds. Plants are divided into two types—dicots, which produce two seed leaves, left, and monocots, right, which produce only one seed leaf.

For herbaceous perennials, vegetative propagation is more important but seed is also used extensively. Seed is a popular method for alpine and rock garden plants but many of these are also increased by cuttings or division. Most trees are propagated by seed. Shrubs are raised by both seed and vegetative methods.

Practically all our hardy fruits are multiplied vegetatively while the vegetable garden supplies an example of the almost exclusive use of seed. Greenhouse plants are increased by seed and by vegetative methods.

Pampas grass, like all other grasses, has abandoned the evolutionary advanced insect pollination for wind pollination, a primitive method also used by the evolutionarily less advanced conifers in the background.

9

2
Conditions and equipment needed

A great many plants, whether increased from seed or cuttings, may be raised in the open. A plot situated in a warm sheltered position is most suitable for this purpose, and a sunny corner of a walled-in garden is ideal. But any plot facing south or south-west and protected from cold winds may be equally good. Ideally the soil should be light and free-working.

A greenhouse, especially if heated, greatly extends the variety and scope of propagation. It is important that a greenhouse admits the maximum amount of light, particularly where most of the propagation is to be done in the short days of late winter and early spring. Good light is essential for the sturdy growth of seedlings.

An unheated greenhouse has its uses but a garden frame can fill the same role more economically. Frames enable the propagator to provide 'close' humid conditions and are best placed in a sheltered shady position. Sometimes frames in the open are provided with a double cover consisting of a sheet of polythene over which one or more glass 'lights' are placed, according to the length of the frame. So-called 'Dutch lights' are excellent for this purpose.

Cold frames in the open have many uses in propagation. Thus, soft wood cuttings from herbaceous plants and shrubs can be raised in them during the warmer months. In the autumn semi-mature or hardwood cuttings such as conifers may be planted in cold frames. Frames are also useful for

Greenhouses considerably extend the range of propagating activities one can undertake. They are essential if one wants to use sophisticated mist propagating equipment.

If one does not have a greenhouse, frames are almost as useful, especially if one makes use of modern techniques like coaxial soil and air-warming cables.

starting early vegetable crops and other plants from seed.

A frame may also be incorporated on a greenhouse bench in an unheated or heated house. Where electricity can be provided for heating, the benefits of bench or frame warming can be enjoyed. This 'bottom heat' is an advantage both for cuttings and in seed propagation. An insulated cable laid in sand on the bench base fulfils this purpose, but adequate for most needs is a purpose-made propagating unit chosen from a wide range on the market. In the simplest units, heat is obtained from a light bulb, above which is a standard plastic seed tray. Such self-contained propagators can be placed in a room window, though this has the drawback of light from one direction only.

Square peat pots fit neatly into ordinary seed boxes and have the advantage that plants can be transplanted with virtually no disturbance to the roots.

The seed tray is the most useful item on the propagator's list. Standard wood or plastic trays are 14 in long by 9 in wide and 2½ or 3 in deep, and may be used for seed raising or rooting cuttings. Boxes in other dimensions are available and may prove to make better use of space in some cases.

Plastic pots come in a wide range of sizes and are preferable to clay pots. Sizes specified represent the diameter across the top; 3½ and 4½ in size pots are generally most suitable for propagation. Half-pots and pans are available in equivalent sizes. Boxes are used to raise moderate or large numbers of seedlings or cuttings, such as annual flowers and chrysanthemums. Pots and pans are ideal for small quantities of either cuttings or seed.

Modern plastic propagating units like this one are easy to use. The hot-plate underneath provides constant mild warmth and the case itself is covered with a clear plastic dome.

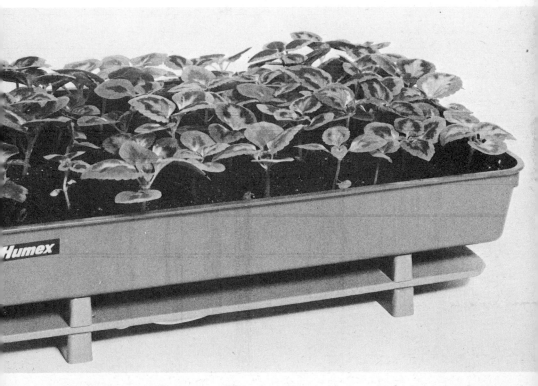

Peat pots, either single or in strips, extend the range of containers for seeds or cuttings. These are planted complete, and subsequently decompose, reducing the transplanting 'check' to young roots.

Equipment used for various operations includes most of the ordinary garden implements, such as spade, fork, rake and hoe. A trowel is very necessary for planting, and one of the most important tools is the dibber. Pencil-thickness is a useful size for greenhouse work while larger dibbers for use outdoors can be about 1 inch in diameter.

A good knife is traditionally the propagator's best friend: strongly made, sharp, and with a handle that provides a good grip. A special knife is used for budding. Secateurs (pruning shears) are also needed and a fine sieve for sowing, while patters are used to level and firm the compost in containers. A pot bottom serves for firming the soil in pots. A watering can with a fine rose is essential, and a syringe with choice of nozzles is serviceable. Last, but of equal importance, comes a good supply of labels, plastic or metal being more durable than wood.

3
Saving seed

Can the gardener save his own seeds and get good results from them? This is possible but there are complications, and with some kinds of plant it is definitely inadvisable. The chief difficulties are as follows. In many districts it is impossible to secure proper ripening and drying owing to moist weather in the autumn. There is always the risk of indiscriminate cross-pollination giving rise to worthless hybrids.

In the case of vegetables, it is advisable to restrict home seed-saving to those that ripen their seed early and are relatively easy to harvest. Such crops include peas, broad beans, runner beans, French beans and tomatoes.

Many shrubs are true species and will breed true from seed. Examples are several brooms, tree lupins, berberis species (excluding *B. stenophylla*), cotoneasters and *Eccremocarpus scaber*. Saving seed from alpine plants is both interesting and successful as many of these are also true species and are easy to raise from seed.

The innumerable hybrids of herbaceous plants cannot be reproduced true to type from seed even if self-pollinated. Excellent seedlings can, however, be raised from many of these plants in a variety of colours. Examples are dahlias, lupins and delphiniums. Plants chosen for seed collection may require careful watching as the seed is often discharged on ripening. Brooms, laburnum and tree lupins are examples. Many shrubs such as berberis, cotoneasters and the snowberry produce their seeds inside berries. When extraction is done it is usually advisable to sow at once.

Further examples of herbaceous perennials from which plentiful seed can be collected–though results may be variable–are sidalcea platycodon, campanulas, hemerocallis, alstroemeria, anthericum, oenothera, various poppies and verbascum. Coreopsis and gaillardia are normally raised from seed and are best treated as biennials.

Seed of various annuals, such as clarkia and larkspur, may be saved, but there is always a risk of cross-pollination and deterioration of the stock. Sweet pea seed is often home-saved successfully. Quite good results may be secured with seed saved from various biennials, such as Sweet-William, wallflowers and Canterbury bells.

Many seeds like those of the yew shown here are designed to pass through the digestive tract of an animal or bird before they will germinate. In practice they often need treating with acid before the seed will germinate.

16

With the exception of seed from certain succulent fruits including certain berried shrubs, all seeds should be thoroughly dried. This may be accomplished by spreading them on sheets of paper laid in shallow containers. A warm greenhouse is an excellent place for drying seed, but sunny windows or airy sheds will also serve. When dried, each lot should be carefully cleaned. A fairly fine sieve is useful for this purpose, and this, together with the simple method of gently blowing on the seed, gets rid of most waste material. Be sure to label each batch when it is collected and write this on the packet or envelope in which the seed will go when dry and clean. Packets should be left unsealed to allow free aeration. The manner of storing seed greatly affects its length of life–ideal is a cool, dry, gently ventilated place.

The colourful berries of many viburnums contain numerous seeds. Stratification is one of the simplest ways of persuading these seeds to germinate.

4
Seed germination

Germination is brought about by allowing seed to have moisture, air and warmth. Perhaps the first is the most obvious requirement since dryness is associated with dormancy, and absorption of water by the seed is a necessary preliminary to germination. Some seeds, such as peas, beans, beetroot and carrots may be soaked in water for a period before sowing. This enables them to absorb water more rapidly than they could in the soil, so germination is accelerated. Most seed, however, requires only a limited steady supply of moisture, such as is present in moist soil, and is likely to be injured by being steeped in water.

The importance of oxygen for germinating seed is not always fully appreciated. From a practical point of view, the gardener must avoid burying seeds so deeply in the soil that air is not freely available to them. Entry of air may also be restricted in wet sticky soil, while heavy overhead watering may cause the soil to run together or cake on the surface, practically excluding air. Under such conditions the seed may be killed or give rise to poor weak seedlings.

Seed sown outside in early spring germinates slowly owing to the coldness of the soil and air. Some seeds, marrow for example, will very likely not germinate at all for they require a minimum temperature of about 55°F before they can germinate. On the other hand, mustard seed will sprout slowly when the temperature is near freezing point. This indicates the wide variation that exists. Up to a certain limit increasing the temperature accelerates germination; its effect is very obvious when seed is sown in a warm greenhouse. Each kind of seed, however, germinates best at a certain temperature, called the optimum. In general, a temperature about 10°F higher than the optimum for normal growth of a plant is most suitable for seed of the same kind.

Seed should not be sown outside until the soil is warm enough to allow germination within a reasonable time. If, for instance, beet seed is sown in February it may decay before germination can occur. Other seeds (broad beans, parsnips and lettuce) may germinate successfully at that time. In the greenhouse, the temperature can be regulated to suit the particular kind of seed.

The germination of most seeds is not affected by light, but

The germination of seedlings in Jiffy strips. The strips tear apart into individual pots for planting out.

light is a vital necessity to all seedlings immediately their leaves are spread out above the soil. If it is absent or not of sufficient intensity, seedlings remain pale and rapidly become weak and drawn up. For strong sturdy growth good light is essential.

Seeds received in the spring should be sown at once in trays or pans, placed in a greenhouse, frame or room. Mild heat is usually advantageous. Spring is the natural time for sowing many kinds and, if germination occurs within a reasonable period, conditions are favourable for the growth of the seedlings. If early germination does not take place the seed should be exposed to cold outside for a period.

When freshly harvested seed is available in the autumn, a portion of it should be sown at once and the remainder in the spring. The autumn-sown batch should be kept indoors until about the New Year, when, if there is no sign of germination it should then be given cold conditions.

Many annuals like these dwarf African marigolds can be grown on until in flower before planting out if sown in peat pots.

During the waiting period, frequent observation is necessary for signs of germination, and unless one has some knowledge of the usual time a particular kind requires for germination, no seed should be discarded for at least 2 years.

5
Seed sowing in the open

Most of our vegetables, many trees and shrubs, various flowering annuals biennials and perennials are raised by sowing seed in the open ground. Sometimes the seed is sown where the plants are to remain and grow to maturity, in other circumstances the object is to produce seedlings for transplanting to another site. In the latter case, particular attention can be given to the selection of a good position for the seed-bed.

Soil preparation should be commenced well in advance of sowing, for the improving effect of frost on land turned over by the spade is better than any amount of cultivation. A second phase in cultivation is done shortly before seeding. This is to thoroughly pulverise the soil with a fork or with one of the small power-driven rotary cultivators available. Finally the surface is finished off fine and level with a rake, free from large stones and clods. Remove deep rooted weeds as the opportunity arises.

A very rich soil is not required for seedlings as it induces soft lush growth which is not desirable. Heavy applications of manure are not recommended unless the soil is very poor. Peat, which improves the structure of the soil and its water-holding capacity without unduly enriching it, is beneficial.

A suitable fertilizer mixture for seed-beds is as follows:
    4 parts superphosphate

1 part sulphate of potash

4 parts bone meal

This is applied at the rate of 2 to 3 oz per sq yd. On rich soil, the hoof and horn could be reduced or omitted. Alternatively, a compound fertilizer with an analysis of 6% N, 8% P_2O_5 and 6% K_2O could be used at the same rate. Lime is also important in the raising of seedlings. As a general rule, $1/4$ lb of hydrated lime (slaked lime) or 6 to 8 oz ground limestone per sq yd should be beneficial to most soils. Rake this into the surface before sowing.

There are two methods of open ground seeding–a) *sowing in drills*, and b) *broadcasting*.

Sowing in drills is preferable for most seeds and is far more often used. It ensures better covering of the seed and allows for more convenient weeding and surface cultivation among the seedlings. Drills are prepared by the use of a draw hoe, triangular hoe, and with the aid of a strong garden line.

Stretch the line tightly from one end of the plot to the other. Take particular care to get the first drill running straight and not at a slant. The corner of the blade is drawn through the soil close against the line, the operator walking backwards, preferably on the line to prevent it being moved by the hoe.

For even germination, the drills must be level at the bottom, and their depth will depend on the type of seed to be sown. Similarly, the distance between the rows varies according to the crop, and may be from 6 to 18 in, but 12 to 15 in is most usual.

Sowing is done by holding a small quantity of seed in the hand and allowing it to fall in a steady even trickle by the movement of the fingers and thumb as the hand moves forward. Thin sowing is usually advised, and seeds should certainly not touch each other. A method of making seed go further is to sow a few seeds at intervals in the drills at the approximate distance apart the seedlings should stand. After sowing, fine soil is drawn over the seed with the back of the rake and the surface made level.

A small seed drill is a useful implement in gardens where there is a fair amount of sowing to be done. This little hand-pushed machine draws the drills, distributes the seed and sometimes covers it, all in one operation. It enables sowing to be done more quickly and more evenly than is possible by hand, especially when the sower is inexperienced.

Broadcasting is used only for a few kinds of seed, such as hardy annuals sown where they are to flower, and grass seed

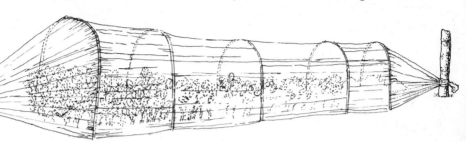

Polythene tunnels are easily constructed and form an ideal way of getting vegetable seeds off to an early start.

for a lawn. Shortly before sowing, the ground is raked absolutely level and very fine. Finally draw the rake firmly through the soil so as to leave the surface as a continuous series of small drills. Over these the seed is scattered thinly and evenly by hand and is then covered by raking in several directions. Grass seed for lawns should, however, be sown very thickly to ensure a good sward.

To provide better cover it is often an advantage to sieve or scatter by hand a little moist soil and peat over the sown area. This is particularly valuable if the seed-bed is somewhat wet and sticky; in such circumstances it may be advisable to cover seed sown in drills similarly.

The correct depth to cover seed depends upon their size. Fine dust-like seed such as poppy should hardly be covered at all; turnips, carrots, leeks and onion require about 1 in cover; radish, parsnip and beet $1^{1}/_{2}$ in; peas and lupins 2 to $2^{1}/_{2}$ in, and runner beans 3 in.

In early spring, rather shallow sowing may accelerate germination as the surface soil warms up first, but in summer, deeper sowing may be advisable to ensure a reliable moisture supply. Sowing too deeply, however, may starve the germinating seed of oxygen as well as hindering and delaying the seedlings in reaching the surface. After sowing, it is usual to

Sowing pelleted seed. This can confidently be sown at the correct spacing at which one wants the plants to grow.

firm the soil over the seed unless the ground is wet. This brings the seed in closer contact with the soil particles and the moisture that surrounds them.

Watering the seed-bed after sowing is rarely advisable as it may wash the soil off the seed and also causes caking of the surface. If the ground is very dry, a good soaking should be given before sowing, and watering is often advantageous when the seedlings are well through. Always give enough water to moisten the top few inches and apply with the finest possible spray.

Pelleted seed is seed coated with a layer of inert substance, thereby increasing size and facilitating handling. The spacing of pelleted seed is easy, and treated vegetable and hardy flower seeds can be conveniently spaced in the open with a saving of seed and effort. Similarly under glass, treated seed

Sowing unpelleted seed by traditional methods. Seedlings will later need thinning.

can be easily spaced in trays so that pricking-off is obviated. A range of such seeds is available, including small flower seeds like alyssum and antirrhinums. The pellets tend to absorb moisture and should therefore be kept in a tightly-sealed plastic container. Once sown, it is important to maintain the surface soil in moist condition.

When seed is not spaced out, the seedlings usually come up too closely together and must be either thinned out or transplanted sufficiently to allow maximum development. Some kinds, such as root vegetables, are difficult to transplant and are sown where they are to remain and mature. Thinning may be done in two operations. On the first occasion, the weakest plants are removed and the others left about twice as thick as required. At the second thinning (where all have survived) alternate plants are pulled.

25

Transplanting should be done as soon as convenient, as small seedlings suffer less check from being lifted than large ones. It is often an advantage to soak the seed-bed the day before lifting and to ease up the seedlings with a fork before pulling.

Transplanting should be done in well prepared soil and preferably in dull, showery weather. Roots should be kept covered during the period the plants are out of the soil. The tools for transplanting are a dibber and trowel and a line to ensure straight rows. A dibber is quite suitable for the speedy

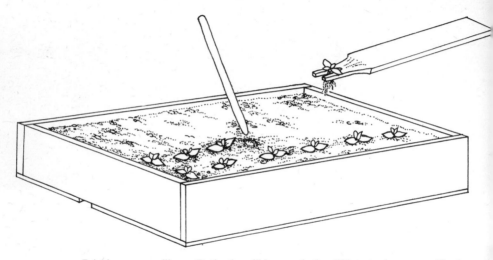

Pricking-on seedlings. Both the dibber and the lifting tool can readily be improvised.

planting of small seedlings, for example those of the cabbage tribe and lettuce. Older plants, however, which have developed a mass of fibrous roots are best planted with the trowel.

To plant with a dibber, it is necessary to hold it in the right hand and the plant in the left. Make a hole with the dibber and place the plant in this, taking care that the roots are not turned upwards. Insert the dibber 1 inch or so from the plant and then use it to press the soil firmly against the stem. When using the trowel, drive it into the ground vertically and pull the soil out towards yourself. Insert the

Jiffys are compressed peat pots contained in a thin polythene mesh. They expand to full size once well soaked.

plant against the back of the hole and replace the soil firmly against it. After transplanting, give a good watering if the soil is dry. Transplanted seedlings also benefit from a light spray on the evenings of hot days.

6
Raising seed under glass

By sowing seed in greenhouses and frames the hazards of weather and pest attack are considerably reduced. The gardener can create his own climate to suit the type of plants he is propagating, and maintain uniform temperatures in a heated structure.

Seed sowing under glass enables a number of vegetables to be raised extra-early in the season. Planted out when the weather is warmer, they result in earlier, higher yielding crops. Examples are lettuces, brassicas (cabbage family) onions and leeks.

Many half-hardy plants must be raised under glass, and are planted outside when the danger of spring frosts is over and growing conditions are more favourable. Marrows, tomatoes and many half-hardy annual flowers provide examples. True greenhouse plants must, of course, be propagated in a heated greenhouse. Many seeds may be germinated in a warm room, but they must be moved to a well-lighted window immediately they are through the soil.

While plants raised under cover escape the ravages of many pests, danger lurks even in a greenhouse. Infestations can spread quickly in a greenhouse and strict hygiene is most essential. All structures should be washed down annually and thoroughly sprayed with a good disinfectant. Leave no rubbish lying about to provide cover. Routine fumigation will keep down such common pests as greenfly and thrips. Water

28

A modern propagating case showing how the soil-warming cables are bedded in sand. Cuttings and seedlings are grown in pots bedded in peat above the sand.

can also be a means of infecting plants with disease, and great care should be taken to ensure a clean supply.

In the greenhouse, seedlings are best raised in a special rooting medium or growing mix (compost).

In the US, a soilless mix consisting of equal parts milled sphagnum peat moss, fine horticultural vermiculite and small granular perlite (or sand) has been used with considerable success.

The advantage of soilless mixes is a high degree of

uniformity because the components are not variable as is the case with loam. They are also lighter in weight, more convenient to store, and unlike loam, do not require sterilization. The main disadvantage is that these mixes need great care in watering otherwise, if overwatered, they will produce massive crops of liverworts: if underwatered and allowed to dry out they are almost impossible to re-wet. They also need regular feeding with liquid feeds because they contain few plant foods themselves.

Quite a few soilless mediums are available under a number of brand names. Be sure to select a brand that is available in both seed and potting because it is really asking a lot of one

Another view of the propagator shown on the previous page. The glass panels slide to allow for ventilation and the control box at the end determines air temperature
Right: A greenhouse given over mainly to plant propagation.

mixture to serve both purposes adequately.

Plastic seed trays and pots should be thoroughly washed and dried before use. Wood trays and clay pots are more difficult to keep in an ideal hygienic condition, which is why plastic receptacles are recommended. It was once thought necessary to put drainage 'crocks' in the bottom of pots and trays before adding the compost, but this is no longer done by commercial nurseries and no loss of quality has been noted in the plants.

So fill receptacles to the brim with compost then firm with gentle finger pressure. The final level surface should be about an inch below the rim. With small seed in a pot, finish off by sieving a little fine compost over the top. With begonia seed, it is a good idea to dust a little fine silver sand over the surface before sowing so that the darker seed can be seen and even distribution ensured.

Use the base of another pot to level the surface in a pot of compost. When firming trays, pay special attention to the corner. Finish off with a home-made wood patter. Before sowing, each pot or tray should be soaked with water from a fine rose or spray. Leave to drain and in the meantime write the labels for each batch of seed with full name and date of sowing and insert in their appropriate receptacles.

Sow using the finger and thumb in preference to direct sprinkling from the seed packet. Aim to sow evenly and thinly. Thick sowing gives rise to a mass of weak seedlings, most at risk from damping-off disease. Covering is done by sifting a little compost over the seed, the depth depending on the type of seed. About $1/8$ in is sufficient for many seeds about the size of lettuce seed. Larger ones like tomato are covered to a depth of about $1/4$ in. It is not advisable to use very fine soil for covering large seed; tomato seed is best covered with soil compost sifted through a $1/4$ in sieve; lettuce and onions covered through an $1/8$ in sieve. Very fine soil tends to cake on the surface thus excluding air. After covering, gently firm again with the patter.

When sowing is completed, it is usual to place all the containers together and cover them with sheets of glass and paper. This is to prevent drying out. Examine daily for signs of germination and at the same time wipe condensed moisture from the glass. Immediately germination occurs the seedlings must be placed in full light. Seedlings grown in shade soon become weak and drawn up. On the other hand young plants under glass may be severely damaged by a period of bright sunshine. The gardener must beware of such a possibility and have temporary shading on hand.

If seeds are sown on moist compost and germination occurs within a reasonable time, watering should not be necessary until the seedlings are above the soil. If the compost dries before germination, the container should be carefully dipped in water and allowed to soak up moisture. After germination, water can be given when required using a can with a fine rose or a spray, and in hot weather this may be necessary at least once a day.

The next stage is to transplant the seedlings or, as it is usually called, pricking-off. Seed trays may be used for this purpose, filled as described earlier. As a rule the richer potting compost is used. Stronger growing plants, such as tomatoes, are usually transferred from the seed tray singly into small pots.

Transplanting seedlings is necessary to allow them space for development, but it does cause a check to growth, hence the fewer moves a seedling has the better. Experiments have shown that the younger a seedling is, the less check it suffers on pricking-off. So this should be done as soon as the seedlings can be handled. Tomatoes, for instance, can be transplanted the second day after they appear above the soil.

Lift seedlings carefully in order to reduce root damage to the minimum. Average spacing in the tray is $1^{1}/_{2}$ to 2 in but 3 in may be needed for the strong-growing kinds. Space them evenly in straight lines. After pricking-off, seedlings need rather warm and humid conditions and may require shading for a day or two. As a rule, hardy and half-hardy plants raised in the manner described are planted outside directly from the trays or pots after being hardened off.

With true greenhouse plants the next move is usually singly into pots or from small pots to larger ones. With each move a richer compost can be used.

As already explained, a cold frame often serves as an intermediate stage between the heated greenhouse and the open garden. Use a frame also for starting early vegetables–like cauliflower, onions and leeks–from seed, as well as raising a number of flowers. Sowing may be done directly in the frame soil or in trays or pans placed in the frame. Unless the frame is heated, do not sow before March. Later sowing usually means better and more rapid development.

7
Annual and biennial flowers

Annual flowering plants are divided into 2 distinct groups: a) **hardy annuals** and b) **half-hardy annuals**. The former are usually sown in the open where they are to flower, while the latter are normally started in a heated greenhouse for planting outside when the weather is congenial.

Hardy annuals, in mild districts, may be sown in the autumn–early September. This should result in earlier and better flowering than with spring sowing. In cold districts, however, autumn sowing is not likely to be successful as the young seedlings may be killed during the winter. Spring sowing is therefore safer, and April–May is the best time.

Annuals that have a relatively good chance of survival from autumn sowing are *Centaurea cyanus* (cornflower), *Iberis* (candytuft), *Delphinium ajacis* (larkspur), *Nigella damascena* (love-in-a-mist) and *Calendula officinalis* (marigold). Among the less hardy annuals are *Clarkia elegans*, *Godetia grandiflora*, *Gypsophila elegans* and *Centaurea moschata* (sweet sultan).

The soil should be in extra good condition. For spring sowing it should be turned over in the autumn and left exposed to the weather. Ensure fertility by application of fertilisers. Phosphates are particularly important for seedlings, and a dusting of superphosphate raked in before sowing is beneficial. Lighten heavy soils by the addition of leaf mould, compost heap material, peat and sand. Prior to sowing, rake the surface fine and level. It is usual to have a border or bed devoted entirely to hardy annuals.

Normally the seed is sown evenly and thinly broadcast; thick sowing may induce damping-off disease. The seed is covered by careful raking but it is usually beneficial in addition to scatter some dry soil over the sown area. This is particularly important on land inclined to be wet and sticky. Large seeds should be sown in drills.

When the plants are well through the soil they must be thinned out to 6 to 9 in apart. Those sown in the summer are thinned out, either in late autumn, or left until the early spring. Thinnings are used to fill in gaps. Weeds must be kept down in the early stages.

Half-hardy annuals are widely used for a summer display, and huge quantities are raised every year in nurseries for sale as bedding plants. However, it is cheaper to raise your own from seed sown between January and March in a heated green-house. Early sowing is necessary with such kinds as antir-rhinums, but many others develop more quickly and may be sown later.

Seed trays should be filled with a sowing mixture and made firm and level on the surface. Water well, and allow to drain before sowing. Scatter seed thinly with the fingers and cover lightly with the same mixture sieved over the surface. Then cover the trays with glass and paper. A temperature of 55° to 60°F is suitable. When the seedlings appear, remove the glass and paper at once and water as required using a fine rose. As soon as possible prick-off the seedlings into other boxes, this time using a potting mixture. Space them so that 54 seedlings go in each standard seed tray.

After pricking-off, keep the young plants warm and humid for a few days to help them to recover from the shift. Some shading may also be necessary. Water so that the soil remains moist.

Gradually, as the plants develop, more ventilation is given, and then in March or early spring the trays are transferred to a cold frame. Here the hardening-off process continues until finally the top of the frame (the 'light') is removed, a week or two in advance of planting out. Do not plant out half-hardy annuals, however, until the risk of frost is past. Examples

36

of this group of flowers are lobelia, petunia and zinnia.

Biennials or flowers treated as such comprise a very useful group for spring and early summer flowering. Seed is usually sown in the open from May until August; early sowing gives the plants a chance to become well established before the winter. Sow in drills drawn 12 in apart. If the soil is dry, water the open drills before sowing.

When the seedlings are large enough to handle they are pricked-off in rows or in beds 6 to 9 in apart. In winter set the plants in their flowering quarters. Examples are Canterbury bells, foxglove, Sweet-William and wallflower.

8
Simple division

Splitting up plants into several smaller ones, each with roots attached, is termed simple division. Plants that lend themselves to this method are those with tufted or matted habit. Alpine plants provide many examples of this type, including aubrietia, arabis, many dianthus and veronicas. Division is also a popular method of increasing perennial flowering plants such as Michaelmas daisies, helianthus, peonies, rudbeckia, doronicum, pyrethrum and scabious. As a rule plants that flower in the spring (doronicum) should be divided in the autumn while autumn flowering subjects are dealt with in the spring. Certain plants, however, are liable to die in the winter if divided in autumn and should be transplanted in spring or after flowering not later than August. These include pyrethrums, scabious, and German irises. Hardy lilies can be divided in spring or autumn. Several hardy herbaceous perennials, such as Michaelmas daisies, do not object to being transplanted during open weather in winter.

In dividing such plants as erigerons, the young portions around the outside should be selected and the older centre discarded. Some plants are difficult to tear apart, but this is facilitated with a handfork. Quite a few can be chopped up neatly with a sharp spade, examples being the yarrows and golden rod. This is a common method in nurseries.

Spacing for many kinds is 1 ft apart each way, but some of the taller growing species, such as aconitums, *Chrysan-*

themum maximum and delphiniums, are planted 2 ft by 18 in. The plants should be made quite firm.

Some of the shrubs which may be increased by simple division are *Kerria japonica*, several of the berberis such as *B. stenophylla*, ericas and spiraeas. These may be lifted, divided and replanted in spring or autumn. Any shoot with roots attached will soon make a new plant. Sometimes a method called 'dropping' is used with shrubs intended for division. This consists in lifting the shrub and taking out some soil beneath it so that the plant is dropped 4 to 6 in. This induces the stems to produce roots from their bases thus facilitating division.

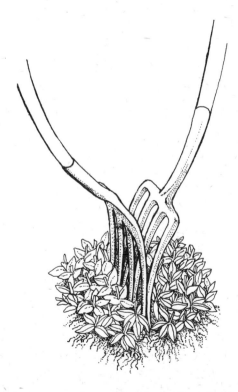

Division of hardy perennials is a simple matter of placing two forks back to back and pulling the plant apart in the middle.

39

9
New plants from old stems

Quite a number of plants produce stems that are capable of giving rise naturally to new plants. These stems have been modified or adapted for this particular function.

Runners. These are slender stems which grow from the parent plant along the ground. The roots and shoots of a new plant arise at each node and soon become established in the soil. The decay of the internode severs the connection with the parent plant. Sometimes the gardener assists the runners of strawberries to root by pegging them down in the soil or by placing a stone on the runner near each node (junction of leaf and stem). The offset is similar to the runner, and is found in the house leek. In this case a slender stem grows out from the crown, bearing at its tip a fleshy rosette of leaves which is capable of taking root at its base.

Suckers. The roots of raspberry spread in the soil a considerable distance from the parent cane. Often they produce buds at various points which develop into shoots coming above the ground to form new plants. These are called suckers and are the principal means of increasing raspberries, apart from raising new varieties by seed. Other plants which produce suckers are plums, lilacs and filberts.

Rhizomes. The swollen structures of the Solomon's Seal and the German iris which are found at the soil surface are also modified stems. Each with a bud attached will grow into a new plant. These swollen rhizomes are also storehouses of food for the young plant. Some plants, however, such as

A sequence showing the propagation of sansieveria by division.

grasses and sedges, produce slender rhizomes like strawberry runners, but as they are underground they are often erroneously mistaken for roots. If such rhizomes are broken into pieces, each with one or more nodes may produce a new plant.

Tubers. Another type of underground stem is the tuber which is very convenient for reproduction. The potato is a common example and each of the buds or 'eyes' it possesses is capable of producing a shoot and roots. Sometimes tubers are cut in pieces and each piece having an eye may give rise to a plant. Other plants that produce tubers are Jerusalem artichokes and tuberous begonias. A clear distinction should be made between tubers and plants with tuberous roots, such as dahlias. The latter are not tubers and can only be used for propagation if there is a 'heel' of true stem attached having one or more nodes or eyes.

The tubers of such garden plants as potatoes, Jerusalem artichokes and begonias are stored over winter and planted in the spring. The latter keep well if bedded in dry sawdust or dry peat moss and kept in a frostproof store. Begonias are usually started into growth in heat.

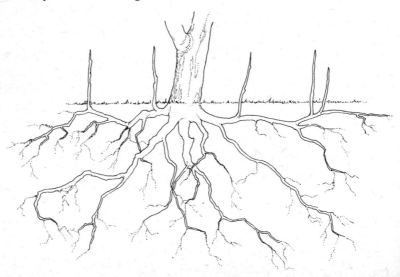

Many woody plants can be increased by suckers. These are best detached, potted up and grown on before planting out.

10
Bulbs and corms

Some of our best garden plants are reproduced from bulbs and corms. These are also classified by botanists as modified stems. Corms are solid structures with one or more buds on the topmost side. When the corm is planted the buds grow upwards and produce foliage and flowers. In doing so the food contained in the corm is used up and the old corm shrivels. As the plants grow, however, a new corm is formed at the base of each shoot. Thus as many new corms may develop as there are buds on the old. Gladioli and crocus are good examples of corms.

In some species of this type (gladioli) large numbers of new buds arise on the old corm and develop into small corms about the size of peas, called spawn. These will grow into ordinary corms but take two or three years to reach flowering size. Spawn may be induced to arise by the artificial method of making two or three cuts across the base of the corm.

A bulb has a more complicated structure than a corm and can be observed by cutting a tulip lengthwise. This is seen to consist of a short, thickened stem bearing roots on the underside and thick fleshy leaves on the upper side which encircle each other. Right in the centre and enclosed by the leaves is a large bud consisting of the undeveloped flower and foliage. Smaller buds may also be found on the short stem between the fleshy leaves.

Under suitable conditions, the central bud grows upwards

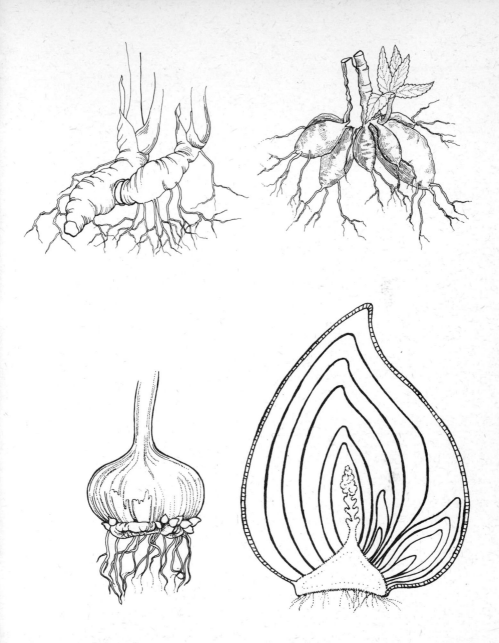

Above left, a typical rhizome–an iris. Above right, a typical tuber–a dahlia. Below left, a typical corm–a gladiolus. Below right, a typical bulb–a daffodil. Left: Mahonia, like many shrubs, can be increased by layers, cuttings, offsets or seed.

45

and produces flowers and foliage. Small buds if present also develop, but few of them produce flowers. All the buds give rise to new bulbs. A large bulb usually produces several daughter bulbs but a small bulb may produce one only, which however is normally larger than its parent. This sequence of events is repeated year after year so that the number of bulbs steadily increases. Plants with scaly bulbs, such as lilies, are readily propagated by simply breaking off the scales and inserting them in a sandy compost.

Certain bulbous plants produce small or secondary bulbs called bulbils. For instance, on the flowering stalks of garlic and tree onions, these are found instead of flowers. Each is capable of developing into an ordinary plant. Bulbils are also found on the stems of some lilies such as the tiger lily, *Lilium tigrinum* and the golden-rayed lily of Japan, *Lilium auratum*. With hyacinths, bulbils may develop on cut or broken bulb scales.

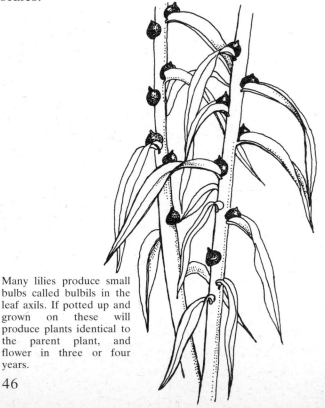

Many lilies produce small bulbs called bulbils in the leaf axils. If potted up and grown on these will produce plants identical to the parent plant, and flower in three or four years.

46

11
Cuttings

Any portion of a plant, whether from the root, the stems, or the leaf, when separated from the parent plant for propagation purposes, is usually described as a cutting.

The type of cutting, including its degree of maturity and the time and manner of securing it, affect propagation. As a general rule, cuttings made from young growths of the current year are preferable. They should always be secured with a sharp knife so as to give a clean cut which promotes rapid healing. Jagged wounds and mangled tissue may cause decay before rooting can take place.

Conditions needed

The essential conditions for rooting cuttings are the presence of moisture, air, and a certain degree of warmth. The need for moisture is obvious seeing that the cutting has been isolated from its former source of supply and is in real danger of being desiccated. This applies particularly to leafy cuttings, as their leaves are still giving off or transpiring moisture. Early insertion, therefore, in a medium such as moist soil, sand or compost, is essential. Moreover, in the case of leafy cuttings loss of moisture must be reduced to a minimum. This is usually achieved by keeping the cuttings in a shaded frame and maintaining a humid atmosphere.

Sand and peat mixed together form an excellent rooting medium because while the sand promotes aeration the peat retains moisture.

Propagation of a variegated rubber plant from leaf bud cuttings. These are taken with a piece of the old stem attached and a new plant will grow from the point where leaf and stem meet.

Rooting of cuttings is usually forwarded by heat, but naturally this varies according to the type of plant. Thus, cuttings from stove plants, such as crotons, require a higher temperature than chrysanthemums or dahlias. An important rule, however, is that the air temperature should be below that of the medium in which the cuttings are inserted.

This means that while rooting is forwarded, the development of the buds and leaves is not so rapid. Experience has proved that rooting should preferably precede the growth of the shoots. This is why bottom heat is often used in the propagation of cuttings.

Another general rule is that soft leafy cuttings respond to higher temperatures than their mature counterparts. Hardwood cuttings usually take a considerable period for callusing and rooting whatever the conditions, so that heat may result only in the immediate growth of the shoots and leaves. This sometimes occurs when hardwood cuttings are planted outside in the spring, and as the cuttings have no roots to make good the loss of moisture from their leaves they may shrivel up and die.

When leafy cuttings are inserted in a frame they usually require shade from bright light to prevent wilting. If shaded continuously, however, the leaves are unable to manufacture food–light is essential for this process–and the cuttings may

48

die of starvation. The correct procedure in such circumstances is to shade for a few days and then gradually reduce the shade until they are fully exposed. Hardwood cuttings do not require light until they begin to produce shoots.

The conditions that are suitable for rooting leafy cuttings, that is, warmth and moisture, are also very favourable to certain plant diseases. Strict hygiene is, therefore, necessary in propagating frames. Although it is necessary to retain as much foliage as possible on the cutting, the leaves that are likely to touch the soil should be cut off. The frame should also be examined frequently and any dead leaves carefully removed.

Effect of Auxins or Hormones

Botanists have shown that all growth and development in plants is controlled mainly by highly active chemicals called 'hormones' or 'auxins'. Research has led to the manufacture of similar substances that are sold as 'rooting hormones'.

For hormone treatment of cuttings by the amateur it is generally advisable to use a powder. Care should be taken to ensure that the powder is of the correct strength for the type of cutting and to read the maker's instructions. As a rule the base of each cutting is moistened before inserting in the powder. Afterwards, try to ensure that the powder is not brushed off when inserting the cuttings.

Stem cuttings

Next to seed, stem cuttings are the most convenient and popular method of propagation. First of all, it is essential for the cutting to have a sufficient reserve of food stored within its tissues to enable it to remain alive until roots and shoots have been produced, when it can secure and manufacture food for itself.

As a general rule, cuttings from young plants root best, but, if older plants are cut back hard, very often they can be induced to produce suitable material.

Broadly speaking, there are three types of stem cuttings, namely hardwood, softwood and semi-hardwood.

Hardwood cuttings are made from the mature stems of shrubs and trees, and provide the simplest of all methods of

50

propagation. Such cuttings are often easy to secure and root readily in ordinary soil in the open. Hardwood cuttings are usually made from firm stout stems of the current year's growths. The length of these cuttings varies from about 4 to 12 in.

It is often advantageous to take hardwood cuttings with a 'heel', that is, with a piece of older wood attached to the base. This is essential in the case of conifers which are propagated from small side-shoots. Cuttings secured without a heel are usually cut just below a node (junction of leaf and stem), but in the case of clematis, internodal cuttings are preferred. For plants with pithy stems, it is advisable to sever the cutting just at the point where the current year's growth originates, as here the pith area is reduced to a minimum. On all cuttings at least one bud near the tip is necessary for extension growth. Buds are not required near the base as roots are produced not from buds, but from inner cambium tissue.

Autumn is probably the best time to secure hardwood cuttings. They may then be planted immediately or tied in bundles and laid in moist soil until the spring. A wide range of deciduous plants are increased by hardwood cuttings including the soft fruits, currants and gooseberries, and many shrubs and trees.

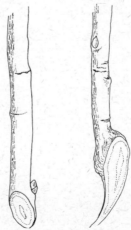

Types of cuttings. Left, an internodal cutting with a bud. Right, a hardwood cutting with a heel.

The 'Plastic wrap' method of propagating fuchsia cuttings. These are placed on a layer of moist sphagnum moss over black polythene.

Cauterizing a poinsettia cutting to stop 'bleeding'. This encourages both callusing and rooting.

52

Another layer of sphagnum is placed over the cuttings, the polythene placed over that and then the whole rolled up and secured with a rubber band.

Pleomele reflexa 'Song of India', for all its beauty, is rare and expensive like any plant that is difficult to propagate.

53

Many cuttings of the hardwood type can be rooted easily in the open garden. Select a sheltered position reasonably free from weeds. The cuttings are planted simply by putting down a line and taking out a shallow trench having a straight vertical side next to the line. Place the cuttings in the trench a few inches apart and at least three-quarters their length deep. The soil is then replaced and trodden in firmly against the stems. If there is more than one row it is usual to allow at least 12 in between them. As previously explained, hardwood cuttings may be planted either in autumn or in spring. Examples of species rooted in this way include blackcurrants, gooseberries, forsythia, ribes and many evergreens such as common laurel (prunus), *Lonicera nitida*, *Viburnum tinus*, veronicas (hebe) and griselinia. Conifer cuttings may strike in warm sandy soil in a sheltered position.

Softwood cuttings have the advantage of being quick-rooting, but they must always be given the protection of some form of glass covering. A great many herbaceous perennials, such as delphiniums, lupins, phlox, dahlias and chrysanthemums, are readily increased in this way. The young shoots produced in the spring are mostly used, each providing a basal cutting. Other miscellaneous plants of a herbaceous type, violas, penstemons and pelargoniums, provide cuttings that may be rooted during the summer and autumn.

Sometimes it is a good idea to cut plants back after flowering to induce the production of cuttings. Aubrietia, alyssum and arabis are examples of plants treated in this way. The important point is to secure young, firm growth, whatever the time of season.

Ornamental shrubs in great variety are readily increased from softwood cuttings, that is cuttings made during the summer, from the immature tips of current growths. Softwood cuttings from shrubs should be reasonably firm and if the shoots snap off cleanly without bending this indicates that they are not too old for propagation.

In preparing softwood cuttings, it is usually recommended to cut below a node, but quite a number of plants can be rooted from internodal cuttings. These include verbenas, antirrhinums, chrysanthemums, hydrangea, fuchsias, heliotrope, calceolaria, penstemons, dahlias, violas and lavenders.

54

Sometimes it is difficult to find the right type of softwood cutting on a plant, but it may be possible to induce suitable growths by cutting back a shoot or pinching out its top.

Softwood cuttings to avoid are thick, coarse, over-vigorous shoots; flowering shoots are also inadvisable but are sometimes used when there is nothing else available. In this case the flowering tip should be cut off.

Softwood cuttings are usually prepared 2 to 4 in long. Any of the lower leaves likely to touch the propagating medium after insertion should be removed. Some trimming of the tips and shortening large leaves is often advisable to prevent wilting, but it is important to leave as much foliage as possible, as this is known to promote rooting.

A special type of softwood cutting used in the propagation of pinks is called a 'piping'. Pipings are secured simply by pulling off the ends of young shoots. The lower leaves are removed and the cutting is ready for insertion. As a rule, softwood cuttings are inserted immediately they are prepared, but a few species, such as the cacti group and zonal pelargoniums, should preferably be laid on the bench to dry before planting.

Practically all softwood cuttings must be given the protection of a glass or plastic covering to allow the maintenance of a moist, close atmosphere which retards the loss of moisture from the cuttings. Rooting is also promoted by warmth particularly when it is applied directly as bottom heat to the propagating medium. In such circumstances softwood cuttings have been known to root in 14 days. The only type of cuttings likely to fail are those with soft woolly leaves such as several alpines which quickly rot in the moist warm atmosphere.

Certain shrubs (cistus and azaleas) root better under conditions that are less close, such as the open staging of a greenhouse. This method is also used for the propagation of chrysanthemums and dahlias. In such circumstances some shading may be necessary for a few days and is best provided with sheets of paper laid lightly over the cuttings.

Cold frames in the open garden are suitable either for softwood cuttings or those made from semi-mature wood. Rooting, however, takes longer than in heated frames, but

A sequence showing the propagation of ivy from tip cuttings. A wire hoop is inserted in the pot to hold the plastic up and away from the leaves of the cutting itself.

Taking a cutting of a croton. The cut should be dipped into powdered charcoal to prevent 'bleeding'.

this may not be of great importance to the amateur. The position of the frame, is important. If it is sited in full sun it will, of course, be warmer than a shaded frame, and this results in quicker rooting. Unless, however, precautions are taken, the cuttings may be scorched and killed if fully exposed to strong sun in the frame.

As a general rule, when a cutting has been prepared it should be inserted in a moist medium as soon as possible. This applies in particular to softwood leafy cuttings which lose water rapidly after they have been isolated.

Various mixtures and composts are used as propagating media, but sharp silver sand is the basis of most of them. Sometimes sand is used alone, and although it is a very good rooting medium, cuttings must be transferred from it into soil immediately they have rooted, as sand contains no plant foods.

Perhaps there is nothing better than a mixture of peat and sand. Suitable proportions are 2 parts sand and 1 part peat by bulk. Such a compost, while being well aerated, is also retentive of moisture. Plants belonging to the heath family do particularly well in a peaty compost, but practically all cuttings may be rooted in it.

There are various other mixtures used for cuttings, the most common being different proportions of sand, peat and soil, such as 3 parts loam, 1 sand and 1 peat; all by bulk. A common practice is to finish off the surface of the soil or compost with a sprinkling of silver sand. This means that when the cuttings are being inserted, some of the sand on the surface trickles to the bottom of the hole where it is in direct contact with the base of the cutting.

In frames, cuttings are either inserted directly into the prepared soil or compost in the frame or into compost contained in receptacles such as boxes or pots, which are placed in the frame. There is a lot to be said for the latter method. It allows cuttings to be transferred from the frames at any time, perhaps for hardening off after rooting, without transplanting them. Pots and pans are ideal for small quantities of cuttings. Moreover, it is well known that difficult species often root best when inserted around the pot sides. No doubt this is due to good aeration.

Whether the compost is contained in receptacles or in frames, make it reasonably firm, and ensure that it is nicely moist when the cuttings are being planted. Insertion is done with a small dibber in rows a few inches apart, the cuttings almost touching one another, and each being made firm. Afterwards give a good watering to settle the compost around the cuttings.

The after-care of cuttings in frames consists in frequent inspections, particularly where the frames are heated. Usually these are examined every morning; water if necessary, remove cast-off leaves and wipe off the condensed moisture from the glass. Cuttings in cold frames should also be kept free from dead leaves, and should be watered sufficiently often to keep the compost just moist but not wet. Shade should be given, particularly during the first few days, to prevent wilting. No ventilation should be given, except to lower a high temperature–over 80°F is high for most hardy cuttings–until there are signs that rooting is beginning. From then onwards cooler and more open conditions should be allowed until the lights are removed altogether.

Of course many cuttings planted in cold frames during the autumn may not rot until the following spring or summer. In such circumstances, the frames should be kept closed during the winter. Ventilation is given and increased gradually in the spring as the weather gets warmer. Some cuttings, such as conifers form a very hard callus layer, which appears to prevent root production. Paring this callus with a knife is usually beneficial in promoting rooting.

Semi-hardwood cuttings are difficult to define and vary from those taken off shoots just beginning to mature at the base to cuttings from shoots almost, but not quite, ripe. The period for securing semi-hardwood cuttings is approximately from early June until the end of September. Shrubs successfully rooted at this period include lilac, hydrangeas, spiraeas, cotoneasters, jasminums, pyracanthas, brooms and rose species. Evergreens including ceanothus, berberis, evergreen azaleas and a number of conifers seem to root particularly well about this time. With most of these the cuttings consist of short lateral growths taken with a heel.

59

The time taken for the various types of cuttings to root is, of course, affected by the conditions under which they are kept after isolation, and depends also on the species. As a general rule the more mature the cutting the longer it takes to root. On the average softwood cuttings take from 2 to 6 weeks, the semi-mature type root in 5 to 25 weeks; hardwood cuttings rarely start rooting in less than 8 weeks and some take as long as 36 weeks.

Leaf cuttings

Buds arise on the leaves of certain plants, making leaf propagation a possibility in such cases. The natural production of buds is very rare, but bryophyllum is an example where this occurs giving rise to new plants. The leaves of certain other plants may be induced to form buds and roots. Broadly speaking, propagation from leaves takes two forms. The first is exemplified by the well-known *Begonia rex* which has large ornamental leaves. In this case the leaf veins are cut through at several points. The leaf is then spread out and pegged down on damp light compost. If enclosed in a warm frame, roots and buds are produced at the incisions, each giving rise to a new plant.

A second type of leaf propagation consists in inserting the leaf stalk into the compost like an ordinary cutting. Strong young leaves are preferred and it is essential for a bud to be present in the leaf angle; otherwise they cannot grow. The leaves are inserted to a depth of about $1^1/_2$ in in peaty compost placed in a well-shaded frame. Plants treated in this way include streptocarpus, saintpaulia and lachenalia. *Begonia rex* may also be treated in this way, but the leaves are cut into strips tapering towards the base, each strip being inserted vertically. Another plant which grows readily from leaf pieces is the house plant sansevieria.

A close heated propagating case is normally necessary to root leaf cuttings.

Root cuttings

Buds appear naturally on the roots of such plants as raspberries and plums and give rise to suckers which are used in propagation. The roots of any such plants are also suitable

60

The propagation of sansevieria by leaf cuttings. Each segment will produce a plant from its base, but the resulting plant will show no variegation until mature.

for root cuttings. Moreover, there are other plants which, although they do not form such buds naturally on their roots, can be induced to do so by ordinary method of propagation. Seakale is a good example of this and most gardeners are only too familiar with the spectacle of pieces of weed roots, like dock and dandelion, sprouting and giving rise to more unwanted plants.

The size of root cutting to take varies with the type of plant, but as a general rule relatively thick roots of a reasonable length are preferred.

Root cuttings. Left, a diagram showing how root cuttings may be taken and, right, how they should be planted.

Amongst fruits, raspberries, and their relations the black-berries and loganberries, are easily increased from root cuttings. Suitable roots are $\frac{1}{4}$ to $\frac{1}{2}$ in thick and are cut in pieces 3 to 4 in long. Such portions may be laid horizontally in a shallow trench and covered with a few inches of fine soil.

Various herbaceous perennials are readily increased by root cuttings. A number of these including *Anchusa azurea* and its varieties, perennial verbascums, eryngiums and oriental poppies may be planted in the open in the same manner as seakale. The ordinary garden phlox *Phlox panicu-lata*, is often increased by root cuttings and this method is quicker than dividing the roots. It consists in selecting the stronger roots and cutting them into pieces about 2 in long. These are sprinkled on a sandy compost contained in boxes or pans and covered with about $\frac{1}{2}$ in of the same medium. Place the receptacles in a frame or glasshouse and, when the shoots are well through, the young plants should be pricked-off in a cold frame or in a nursery bed in the open.

12
Grafting

Grafting is the art of inducing a piece of the stem of one plant, called the scion, to unite with the rooted portion of another, known as the stock. The two, so united, grow together to form one plant, yet each maintains its individuality and a shoot arising from one is distinct from that produced by the other. Normally, however, the role of the stock is to provide the plant's roots while the scion furnishes the top growths.

The primary object of grafting and budding is to reproduce plants that are more difficult or cannot be propagated at all by other vegetative methods. Thus, the various tree fruits, such as apples, pears and plums, are extremely difficult to root from any type of cutting, while by grafting large numbers can be raised with comparative ease.

Another important reason for grafting is to enable certain plants to be grown on roots other than their own, and this is often advantageous. Tree fruits again provide an example and, in the case of apples, several distinct root stocks are used with the object of securing trees suitable for different purposes and conditions.

Roses are usually grafted (or budded), although most varieties root easily from hardwood cuttings.

Grafting has several disadvantages. Grafted plants may be prone to suckering, that is, to producing unwanted shoots from the root stock. This is a bad fault with some plum stocks, and can be a serious nuisance. Usually, stocks for grafting in the open are planted in the soil, and for most subjects such as fruit trees, or roses, should be established at least a year before grafting or budding is attempted.

Grafting out of doors is usually undertaken in early spring just when growth is commencing. The scions, however, should be severed from the parent plant when they are quite dormant. Usually they are then bunched together and laid in damp sand or soil in a shady position.

Success with the operation of grafting is largely a matter of bringing the cambium (an active layer just below the bark) of the stock and the cambium of the scion together, and no union can occur unless this is achieved. This, therefore, should be the aim, and it is facilitated by making firm, clean cuts with a really sharp knife in such a way that the parts fit

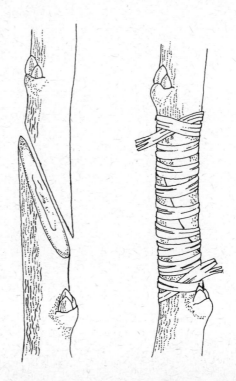

Splice or whip grafting–by far the simplest and most widely used grafting technique of all.

snugly and firmly together.

Scions vary in length according to the type of plant, but a common length is about 6 in. The stocks should be well established and sufficiently strong to sustain the grafted plant.

With most methods of grafting, it is necessary to tie the scion firmly to the stock and raffia is often used for this purpose. After tying, the union is usually sealed around with a wax or jelly to prevent the entry of water and to check drying out of the wounded tissues.

Grafting wax may be purchased ready for use, and some products may be used cold. Petroleum jelly is also used for sealing grafts and is fairly effective. In recent years, adhesive tape is being used both as a tie and to act as a substitute for wax.

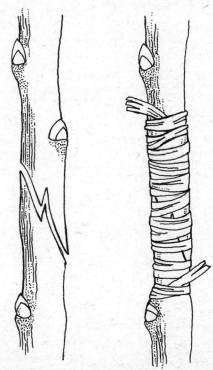

Whip and tongue grafting, a variant on simple whip grafting, ensures an even tighter fit of cambium to cambium.

13
Budding

Although budding is usually regarded as a distinct method of propagation, the principle is exactly the same as in grafting. With the latter method, a complete length of stem is used as the scion, whereas budding consists in grafting a piece of rind, with bud attached, on to another stem or root. The union takes place between the cambium layer attached to the piece of rind and the cambium layer of the stock. The bud is there to provide the first shoot of the plant but takes no part in the actual union. Some of the disadvantages often attributed to grafting, such as suckering and the possibility of using unsuitable stocks, apply of course to budding.

Roses are the most important group of plants that are increased by budding, but it is also widely used in the propagation of fruit trees.

Budding is generally done at the height of the growing season, that is in early summer. This is because during these months the rind of most plants separates easily from the wood and facilitates the operation. Stock plants established on the site for at least a year are used for this purpose. Before commencing to bud, any laterals near the stock base should be trimmed off so as to have a clean stem for the bud.

Generally, the bud is inserted near the ground, and in the case of roses almost on the root. This tends to obviate the production of suckers and encourages the scion to produce roots also, which is desirable with roses and other plants.

Young stems of the current year's growth provide the

buds. After being severed the stems should not be left in strong sun and other measures such as standing in water may be necessary to prevent drying out. For this reason dull, moist weather is preferable for budding, and the buds should be inserted in the stock immediately they are isolated. Select plump wood buds from medium-sized stems and do not use those on the immature soft tips.

There are several methods of budding but by far the most important is called 'shield' or 'T budding'. This is a simple operation which with a little practice anyone should be able to perform successfully. A good sharp budding knife is essential. The stock is prepared by making a clean vertical cut about 1 in long right down to the wood, but no deeper. At the top of this, make a transverse cut to form a 'T'-shaped incision. A bud is then sliced from the stem with about $\frac{1}{2}$ in rind on either side of it.

There is always a great deal of controversy as to whether or not the wood adhering to the rind should be pulled out. As a rule, it is better to do so because this allows improved cambial contact. In removing the wood, however, there is the risk of pulling out the base of the bud, which is a little green structure about the size of a pinhead. The attached leaf on a bud should be cut off so as to reduce transpiration to the minimum, but the leaf-stalk or petiole is allowed to remain and serve as a handle.

When the bud is ready, the rind is carefully and cleanly

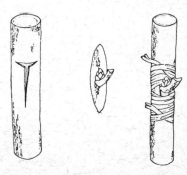

Budding is simply a rather specialised form of grafting. The method shown here is known as 'shield' or 'T-budding'.

lifted from the wood on either side of the incision on the stock. Into this the little shield held by its tiny handle is inserted from the top. Extra care and taking of pains are necessary for the beginner to ensure a neat job and improved prospects of success. Speed comes with experience. To complete the operation, bind the union firmly with raffia, leaving the bud, of course, uncovered. The binding prevents drying out and the entry of rain or insects.

The modern hybrid tea and floribunda roses are almost entirely propagated by budding. The technique is also used for many ornamental cherries.

14
Layering

Layering is a method of plant propagation which is designed to induce plant stems to produce roots while they are still attached to and sustained by the 'parent' plant. It is a reliable means of increase and is often adopted for species which are difficult or impossible to root from cuttings. Layering does not lend itself to the production of large numbers of plants and is, therefore, used by the nurseryman only for the more difficult species.

This method is of particular advantage to the amateur who does not require a large quantity of any particular species. For layering is comparatively easy to perform and does not require the facilities necessary for certain other methods of propagation, such as softwood cuttings. Layering is used largely for increasing woody plants, but a few of the herbaceous type, notably border carnations, are propagated in this way.

The soil for layering around the stock plants should have liberal applications of peat and sand to make it porous and more retentive of moisture. Weeds should be kept down by cultivation as necessary. Autumn is the most suitable time for this type of layering, but the work often continues in open weather throughout the winter.

Ordinary layering

Ordinary layering may consist of no more than bending the shoots downwards and covering the portion of the stem nearest the ground with soil.

The gardener can easily layer a few shoots of any shrub in his garden, provided the stems can be brought down to the ground. It is essential to use young shoots only. Older branches are usually slow to root and rarely develop into well-shaped specimens. A notch, slit, ring or even the paring off of some of the bark on the part to be layered is usually beneficial Secure a good right-angle bend and fix firmly in the soil using a peg if necessary. As explained previously, sand and peat mixed with the soil will promote rooting.

Most shrubs root in less than a year, and if layered during the autumn or winter are often ready for transplanting the following autumn. A layered shoot should, however, be carefully examined before severing and, if insufficiently rooted, should be left for another year.

Practically any woody plant may be increased by layering, but the nurseryman uses it for plants for which it would be difficult by other means. These include magnolias, pieris, rhododendrons, syringa (lilacs), plagianthus, garrya, photinia, mespilus, leucothoe and nothofagus. Certain procumbent or horizontally growing conifers often seen in rock gardens are convenient plants to layer. Various other dwarf shrubs found on the rock garden may be layered *in situ*. These include *Daphne cneorum* and dwarf rhododendrons.

Tip layering

Perhaps the simplest form of layering is tip layering or

Layering is one of the most certain methods of increasing most shrubs. Note that the layer is partly severed before being placed in the soil, and is then staked to prevent movement.

71

A sequence showing the air-layering of a variegated rubber plant. The most important factor is to ensure a water-tight seal at the top and bottom of the black polythene.

72

A photographic sequence showing the layering of a rhododendron. Note that the layer is held in the ground with a wooden peg, and is also tied to a stake to make it grow upright.

growing point layering, which often occurs naturally and is the normal method of increasing blackberries and loganberries. It may also be used for currants and gooseberries. Tip layering consists in bending down firm young shoots in summer and burying their tips in the soil to a depth of 4 to 6 in, or they may be inserted in pots. Rooting usually occurs in a matter of weeks and the new plants may be severed and transplanted in the autumn.

Air layering

In all the various forms of layering described so far, rooting is induced by bringing the stems into contact with the soil or other moist medium. There is one quite distinct type of layering, however, whereby the moist material is brought to the stem, the latter being left in its original position. This is called 'air layering' and is believed to have been used by gardeners for thousands of years. It is a useful method for propagating rare and valuable plants which are difficult or impossible from cuttings, and where ordinary methods of layering are not practicable. Thus, plants with rigid branches borne high above the ground could not be brought down to the soil. Certain greenhouse and house plants, such as *Ficus elastica*, are propagated in this way.

Clean young shoots are best for air layering. They are prepared by making an upward cut about 2 in long at or about the stem centre. Alternatively, the stem may be girdled by removing a ring of bark about $1/2$ in wide. The wound may be dusted with a growth-promoting substance, and a handful of moist sphagnum moss is packed between the cut surfaces and all around it, so as to give complete cover.

In the past, the main difficulty with air layering has been to prevent the moss from drying out. This problem appears now to have been largely overcome by the use of thin polythene film. This is wrapped around the sphagnum moss in such a manner as to leave no opening which would allow evaporation of moisture from the moss. The film is held in position by a piece of tape tied around the stem both above and below the layer.

15
Propagating perennial flowers

Because of their popularity, the detailed propagation of chrysanthemums and dahlias is of interest to a great number of gardeners and will be considered in the first part of this section.

Chrysanthemums. Increase by cuttings is the standard method and the first necessity for success is to secure these only from healthy plants that are true to type and variety. This involves careful inspection during the growing season, and the propagator should familiarise himself with the symptoms of the common pests and diseases of chrysanthemums.

Plants for propagation are carefully labelled after flowering and then cut down to within 6 in of soil level. They are lifted about the end of October and soil washed from the roots to remove soil pests, particularly slugs and their eggs. All leafy growth on the 'stools' (the mat of stems and roots) should be trimmed off. Place the stools in boxes using clean soil or peat and soil to pack around them. Pot plants may be either left in the pots or knocked out and laid in the boxes.

During winter, keep the boxes in sufficient warmth to afford protection against frost. Unheated frames should be covered with mats during severe frost. Only slight heat is necessary to start the plants into growth and the young basal shoots are used as cuttings.

Most suitable time for taking cuttings varies with the type

of chrysanthemum and also with the purpose for which the new plants are intended. Thus cuttings of the large Japanese varieties which are often grown for exhibition are usually secured in December or January in the British Isles. Those which flower from October to December are, as a rule, taken in January or February, while propagation of the early-flowering types is done in February or early March.

In unheated greenhouses or frames, growth may be delayed and one may have to wait until shoots for cuttings are produced. Some varieties, such as the late flowering 'Friendly Rival', throw up few or no basal growths, and stem cuttings rather than stool cuttings must be taken. Experiments have shown both kinds to be equally useful.

The chrysanthemum cutting is one of the easiest to root in pots or boxes. An ideal rooting compost will allow free drainage, and the following mixture is recommended: 3 parts by volume of coarse sand ($1/8$ in) and 2 parts granulated peat. Heat below the containers at about 65°F is recommended, with air temperature about 60°F.

Take cuttings by breaking off (not cutting off) the young shoots, leaving at least three leaves at the base of each shoot so that further shoots will be made for cuttings. Keep cuttings to a standard length of 2 to $2^{1}/_2$ in in winter and $1^{1}/_2$ to 2 in for the rest of the year. Without further preparation—except a hormone powder dip if desired to promote more rapid rooting—insert the cuttings and water in well. Subsequently, spray over at intervals to maintain humidity according to weather conditions. A standard seed tray should contain no more than 50 cuttings; in pots they are best inserted around the edge.

Once rooted, the treatment varies according to the proposed method of culture. Early-flowering chrysanthemums are often bedded out in a frame where they are protected from frost until they are planted in the open in late spring. Plants for growing and flowering in large pots are potted first in 3 in pots and later transferred to their final 8 or 10 in pots.

Dahlias, like chrysanthemums, are usually propagated from cuttings, and healthy plants are first selected. Lift the plants when the foliage is blackened by frost, cut off the stems close

76

to ground level and thoroughly dry the tuberous roots. Label each root carefully before storing. Essential storage conditions are dryness and freedom from frost. Surround the roots with soil, sand or peat in a fairly dry stage. Before the roots are stored, examine them carefully, cut away any decayed or damaged portions and sprinkle the cut surface with sulphur. Usually in early January the roots are set closely together in boxes or on greenhouse staging and moist peat or light soil packed around them. Artificial heat is necessary for early cuttings, but they will be produced later in a cold house. The house should be light enough to promote sturdy growth.

The new shoots on dahlias arise, not from the so-called tubers, but from the bases of the old stems. It is an advantage to allow the first growths produced to form four joints and then to cut them at the second joint from the base. This induces further shoots to arise. Sturdy growths are best for cuttings, and thin, lanky shoots should be avoided. Sever the cutting below a joint, remove the lower leaves and insert in moist, sandy compost.

As a rule $3^1/_2$ in pots are most suitable for rooting cuttings and about 6 can be inserted around the edge. Place the pots on the staging of a warm greenhouse or in an open propagating case. Bottom heat is beneficial. At 55 to 60°F, rooting occurs in two to three weeks. Rooted plants are potted in 3 in pots using a freely draining compost. Keep them warm and humid for a few days then gradually expose to cooler conditions.

Usually dahlias are stood in a cold frame for a few weeks prior to planting outside. They are generally set in the open in May, but if conditions are unsuitable for planting they should be potted on into larger pots and kept growing.

Division is one way of propagating dahlias for the gardener who has no heated glass. Tubers are lifted and stored in the usual way, then in April they are divided up using a sharp knife. Make sure that one or more tubers are attached to a piece of stem with a bud. Six plants can often be secured from an average size root. The divided portions are then planted out sufficiently deep to protect them from late frost damage.

Seed offers a third method of increase and, although it is not suitable for the general run of named varieties, some excellent strains that come true to type and colour from seed have been developed. These include the Coltness Gem hybrids and other dwarf bedders as well as the Charm dahlias.

Where a heated greenhouse is available, seed may be sown in gentle heat in February or March. Sow thinly in trays or pans of sowing compost and cover very lightly. Prick-off seedlings when large enough to handle into trays or singly into pots. Subsequent treatment is the same as from cuttings.

Hardy perennial flowers can be increased in various ways, and treatment of the best known types is detailed alphabetically as follows:

Acanthus (Bear's Breeches) Root cuttings are the usual method, but division in the spring is also successful. Seed is another means of increase.

Achillea (Yarrow) Easily increased by division in spring or autumn. Soft cuttings strike readily in spring. The species come true from seed.

Aconitum (Monkshood) Sow seed outside in April or May. Divide in spring or autumn.

Alstroemeria (Peruvian Lily) Divide the roots in spring or sow seed thinly under glass in the autumn or early spring. Seedlings must be handled with great care and disturbed as little as possible.

Anchusa (Bugloss) Easily increased from root cuttings about 4 in long planted in the open or in a cold frame in spring.

Anemone (Wind-flower) The popular *A. hupehensis* (syn. *A. japonica*) and its varieties are usually increased from root cuttings in spring. Division in the autumn is another method, but the rooted pieces should preferably be potted.

Anthemis (Chamomile) Readily propagated from soft cuttings in spring or by division in spring or autumn.

Aquilegia (Columbine) The species can be increased from seed if plants are kept well isolated from one another. Excellent strains in mixed colours can also be raised from seed. Sow in the open in spring. Division of named varieties may be done in spring.

Artemisia (Wormwood) Mainly increased by division.

Astilbe (Goat's Beard) Divide during the dormant period or

78

sow seeds under glass in February.

Bergenia Divide in spring or autumn.

Campanula (Bell Flower) The tall herbaceous types are easily increased by division or spring cuttings. Seed may also be used.

Carnation Border carnations are increased by layering.

Chrysanthemum *C. maximum* is readily increased by division or by spring cuttings.

Coreopsis (Tickseed) The perennial species are raised from seeds sown in the open in spring; or divide in spring or autumn.

Delphinium Propagate from basal cuttings taken with a heel in a slightly heated glasshouse or cold frame. Seed sown in heat in January and planted out in April or May will produce plants flowering in the autumn. Seed may also be sown in spring or early summer in the open.

Dicentra (Bleeding Heart) Divide in spring or take cuttings at the same time.

Digitalis (Foxglove) The perennial species are increased by seed or division.

Doronicum (Leopard's Bane) Easily increased by division in autumn.

Echinops (Globe Thistle) Root cuttings planted in a cold frame in spring are effective; or divide in spring or autumn.

Eremurus Propagate by root division in spring. Seed is also feasible but is often slow to germinate and seedlings take 3 years to reach flowering age.

Erigeron (Flea-bane) Divide in spring or autumn or secure basal cuttings in spring.

Eryngium (Sea Holly) Take root cuttings in spring and divide in spring or autumn.

Gaillardia (Blanket Flower) Name varieties do not come true from seed and should be raised by division in spring or by cuttings taken in August or September and protected over winter under glass.

Geranium (Crane's Bill) Divide the plants in spring or autumn and sow seed in spring.

Geum (Avens) The hybrids should be increased by spring division, but several good strains can be raised from seed sown in spring.

Gypsophila (Chalk Plant) *G. paniculata* is easily raised from seed sown in heat in February or out of doors in April. Varieties must be raised either from soft cuttings secured in spring from plants grown in heat or by root-grafting on seedlings. The varieties Bristol Fairy and *flore plena* (double form) are increased in this manner.

Helenium (Sneezeweed) Can be divided almost any time during the dormant period and is also easily increased from spring cuttings.

Helianthus (Sunflower) Increase is similar to *Helenium*.

Heliopsis Easily increased by division.

Helleborus (Christmas Rose) Divide in spring or autumn. Sow seed immediately it is ripe in a cold frame, or in the open.

Hemerocallis (Day Lily) The clumps may be divided in spring or autumn.

Heuchera (Alum-root) Divide in spring or increase by seeds sown in spring. Seedlings, however, are often variable.

Incarvillea Increased by division or seed, which takes 3 years to reach flowering size.

Iris (Flag) Seed is an important means of increasing this large genus, and most species can be so raised. Usually seed is sown when ripe and, after exposure to winter cold, germinates readily in spring in mild heat. Division is also widely used and is the only means of increasing varieties. The bearded irises should be divided immediately after flowering.

Kniphofia The red-hot pokers are normally increased by division in spring. The species come true from seed sown in spring.

Liatris Easily increased from offsets secured and transplanted in spring. Seed may also be used, sown in spring.

Lupinus Lupins are easily increased from seed sown in heat in early spring or in the open later in the season. Named varieties are best increased from heeled cuttings secured when available in spring and inserted in pots. Lupins are difficult to divide, although this may be attempted.

Lychnis (Campion) Easily increased by division in spring or autumn. Seed sown outside in spring provides another method.

Lysimachia (Loose-strife) Propagate by division in spring or autumn.

80

Lythrum (Purple Loose-strife) Propagation is by division.

Malva (Mallow) Easily increased by seed or cuttings.

Monarda (Bee Balm) Divide in spring or use soft basal cuttings.

Montbretia (Tritonia) Easily increased by division.

Nepeta (Catmint) Readily increased by division or by soft cuttings taken in summer or autumn.

Oenothera (Evening Primrose) Insert cuttings of the perennial species in a frame before flowering. Divide in spring. Seed is another method.

Paeonia (Peony) The species should be raised from seed sown when ripe in cold frame. Sometimes germination is slow. Divide in autumn, ensuring that each piece of root has an 'eye'.

Papaver (Poppy) The well-known oriental poppies are easily increased from root cuttings about 4 in long secured and planted in the open in spring.

Phlox *P. paniculata* can be raised from seed, but its many varieties are increased by basal cuttings or by division in spring or autumn.

Physalis (Chinese Lantern) Natural increase by underground stems occurs rapidly.

Physostegia (False Dragon-head) Sow seed in cold frame in spring or divide roots in autumn or spring.

Platycodon (Chinese Bell-flower) Propagation is by seed or division.

Polemonium (Jacob's Ladder) Divide in spring or autumn. Seed may be sown in spring.

Polygonatum (Solomon's Seal) Propagate by division in spring or autumn.

Potentilla (Cinquefoil) Increase by seed sown in the open in spring or divide in spring or autumn.

Pyrethrum The pyrethrum is really a chrysanthemum. It is increased by division in spring or immediately after flowering.

Rudbeckia (Cone Flower) Divide in spring or autumn, or basal cuttings root readily in a cold frame in spring. Seed may be sown out of doors in spring.

Salvia (Sage) Division in spring or seed sown in spring are the usual methods.

Scabiosa Divide in spring only. Seed is another method but varieties do not come true.

Sidalcea Easily increased by division in spring or autumn.

Solidago (Golden Rod) Easily increased by division in spring or autumn.

Stachys Increase by division in autumn or spring.

Stokesia Usually propagated by seed or division.

Thalictrum (Meadow Rue) Divide in spring just as growth commences or sow seed in the open in spring.

Tradescantia (Spider-wort) Basal cuttings in spring root readily or plants may be divided in spring.

Trillium (American Wood Lily) Increased from seed sown when ripe.

Trollius (Globe Flower) Propagate by division just after flowering or sow seed when ripe.

Verbascum (Mullein) Readily increased from root cuttings in spring, or division, or by seed sown in spring.

Veronica (Speedwell) Divide in spring or autumn, or take basal cuttings in spring.

16
Propagation of trees and shrubs

Seed is widely used in the propagation of trees and shrubs. Autumn sowing immediately the seed is ripe is often advantageous. However, as the seed is unlikely to germinate before late spring or early summer, weeds may be a problem. With such seed, therefore, over winter stratification may be preferable. A simple method of doing this is to mix the seed with moist soil, peat or sand or a mixture of these and place in boxes or pots which are buried outside.

In the propagation of shrubs from softwood or semi-hardwood cuttings mist can be used almost without exception.

How the different kinds are increased

Abelia. In July, take half-mature sideshoots with a heel and insert in a close case. Mature sideshoots will also strike planted in a cold frame in November.

Abutilon (Indian Mallow) Soft-wood cuttings root readily in June or July in a close case or sun frame. Ripe wood cuttings may be struck in a cold frame in November. *A. vitifolium* may be raised from seed sown indoors in January.

Acer (Maple) Most of the species and certain varieties, such as *A. palmatum* var. *atropurpureum* may be raised from seed sown in the open as soon as ripe or stratified and then sown. *A. negundo* varieties and *A. palmatum* may be budded on the seedling species. The latter is also grafted under glass.

Acer palmatum 'Dissectum atropurpureum' in autumn colour. The plant is most easily increased by layers, but seeds are occasionally produced. Right, the common houseleek is increased by detaching rosettes and planting them on.

84

Actinidia Cuttings of semi-hardwood will strike in moderate heat. Seed in gentle heat is another method, while layering may be done during the dormant season.

Ampelopsis Softwood cuttings can be struck in a close case in July, and in October hardwood cuttings will root in a cold frame.

Aralia (Angelica Tree) Usually increased from suckers and root-cuttings in spring. Seed is another method.

Aristolochia (Dutchman's Pipe) Softwood cuttings in cold frame in summer will root but slight bottom heat is an advantage.

Arundinaria (Bamboo) Divide old plants in spring and for speedy rooting pot the divisions and keep in a warm humid glasshouse.

Atriplex Propagate by summer cuttings.

Aucuba (Spotted Laurel) Cuttings made from firm shoots 6 to 9 in long and planted in the open during the autumn usually strike readily. Layering is a still more certain method and can be recommended to the amateur.

Berberis (Barberry) Many species come true from seed including *B. darwinii*, *B. gagnepainii* and *B. thunbergii* if each is reasonably isolated from other species. Stratify the seed over the winter and sow in the open in early spring. Special varieties and hybrids such as *B.x. stenophylla* do not come true from seed and must be propagated vegetatively. Cuttings of a large number of both evergreen and deciduous types are made about 6 in long from current season's wood with a heel, and strike easily in a cold frame. Layering and division are other methods.

Buddleia Cuttings are a popular and easy method and will strike in the open. Use ripe wood and make the cuttings 6 to 8 in long. Plant in October or November. Softwood cuttings 3 to 4 in long also strike readily in a close case in July. *b. alternifolia* is often raised from seed sown in heat in February.

Buxus (Box) Most of the boxwoods are easy from cuttings. Take mature side shoots with a heel and plant in a cold frame in September. Division is another method.

Calluna (Ling) Propagate as for *Erica*.

Camellia *C. japonica* (Camellia) is raised from seed sown in heat in February. Prior to sowing it is recommended to soak the seed in warm water for 24 hours. *C. japonica* and *c. reticulata* varieties are usually grafted on seedlings of the ordinary type. This is done in a close case in August. Several varieties including singles may be raised from cuttings of half-mature wood in a close case in July. Another method is leaf-cuttings.

Campsis Increase by seed if available. Other methods are ripe cuttings, suckers and root cuttings in moderate heat.

Caryopteris Small softwood cuttings are secured from plants grown in heat in March and are struck in a close case. Cuttings of similar type secured from outdoor plants in July may also be rooted in a close case. Hardwood cuttings 4 to 6 in long planted in a cold frame in November provide another method.

Ceanothus Sideshoots with a heel are taken in October and rooted in a cold frame. Softwood cuttings will also strike in June in a close case with bottom heat. The evergreen species such as *C. rigidus* are best propagated from mature sideshoots inserted in a cold frame in the autumn.

Ceratostigma (Lead Wort) Soft-wood cuttings with a heel strike readily in a close case. These are secured from outdoor plants in summer or from plants grown in pots which are cut back in the autumn and brought into heat in December or January and forced into growth.

Chaenomeles The well-known *C. lagenaria* (syn. *Cydonia japonica*) Japanese Quince and other species may be raised from seed which is collected when ripe, stored dry over winter and sown in early spring. Varieties are often grafted onto seedling stocks, but plants on their own roots are superior and may be had by layering.

Chamaecyparis (Cypress) The ordinary species such as *C. lawsoniana* (Lawson's Cypress) are raised from seed sown in the open in spring. Special forms and varieties are raised from cuttings consisting of sideshoots with a heel. These are inserted in a cold frame in October. Grafting on seedling stocks under glass is also used to increase the varieties.

Chimonanthus praecox Sow seed in early spring. Layering is another method often used.

Choisya ternata (Mexican Orange) Cuttings of mature wood

4 to 6 in long with a heel will root in a cold frame when inserted in autumn. The immature tips root readily in a close case and are taken in summer.

Cistus (Rock Rose) True species are raised from seed sown under glass in February. Cutting of sideshoots taken with a heel root readily in a close case in summer. Wood that is almost mature may be used for cuttings in autumn and planted in a cold frame.

Clematis Seed may be used when available to increase the species. Grafting is a popular method for the various hybrids *C. vitalba* (Traveller's Joy) is used as a stock, the scions being grafted onto the roots under glass. Internodal cuttings of half-ripened wood taken in August will strike in a close case. They are often inserted singly in pots.

Clerodendon Usually propagated from root cuttings. Fairly thick pieces about 3 in long are secured in winter and may be laid in moist sand until April, when they are planted in outdoor beds.

Clianthus (Glory pea) *C. dampieri*, (syn. of *C. formosus*) is grafted in the seedling stage on seedlings of *Colutea arborescens*. *C. puniceus* is raised from cuttings of half-mature wood about 3 in long, inserted in a close case in June–July.

Colutea (Bladder Senna) The best plants are produced from seed which is kept dry until sowing time in February or March under glass. Half-mature nodal cuttings may be rooted in a close case or frame in July or August.

Corokia May be raised by seed in spring or by cuttings in July or August.

Coronilla (Crown Vetch) Easily raised from spring-sown seed or summer cuttings.

Corylopsis Layering is the usual method but cuttings in late summer will also strike.

Cotoneaster Seed is successful for many species including *C. bullata*, *C. frigida*, *C. lactea* and *C. simonsii*. It should be collected when ripe, stratified during the winter and sown in the open in March. Cuttings are used for several species, such as *C. adpressa*, *C. horizontalis*, *C. microphylla* and *C. rotundifolia*. These should always be taken with a heel and may be either ripe wood taken in November and planted in a

88

Conifers are often propagated commercially by mist propagation. Cuttings kept close are simplest for the amateur.

cold frame, or half-mature sideshoots inserted in a frame in July.

Cupressus (Cypress) *C. macrocarpa* and *C. sempervirens* are raised from seed sown in February or March indoors, or outside later in the spring. Grafting is also done for varieties under glass in summer, *C. macrocarpa* being used as a stock.

Cytisus (Broom) Several species are easily raised from seed, but crossing occurs readily unless they are well isolated. Sow in March or April using pans placed in a cold frame. Cuttings are used for some species and several varieties, e.g., *C. ardoinii*, *C. praecox*, *C. purgans* and *C. purpureus* and its

89

varieties. These may be either half-mature sideshoots with a heel inserted under bell glasses in August or firm sideshoots secured in the autumn and planted in a cold frame. Some types such as *C.x. kewensis* and *C.x. beanii* are grafted onto seedling laburnum in March, in the open.

Daboecia *D. cantabrica* (syn. *D. polifolia*, Irish Heath) may be raised from seed sown in the spring under glass, in pure sifted peat. Cuttings may be struck under bell glasses in October. Division of plants previously 'dropped' is very effective on light, sandy soils.

Daphne Seed is used for several species such as *D. laureola* (Spurge Laurel) and *D. mezereum* (mezereon) and its varieties. Stratify the seed during the winter and sow in early spring in a cold frame. Cuttings are successful for *D. cneorum* (Garland Flower) *D. collina* and *D. retusa*. They consist of semi-mature sideshoots inserted in a cold frame or bell glasses in summer. Root cuttings in heat is the usual method for *D. genkwa*. Another method is grafting under glass in spring. *D. laureola* being the stock for evergreen species and *D. mezereum* for deciduous types.

Deutzia All species and varieties are readily increased by softwood cuttings 4 to 5 in long taken in summer and inserted in a close case or frame. Hardwood cuttings about 8 in long are usually taken with a heel and in mild districts can be planted outside in November. In colder localities they should be inserted in a cold frame and made somewhat shorter. Seed stored dry over winter will germinate in a warm glasshouse in February.

Diervilla (Bush Honeysuckle) Softwood cuttings, 4 to 5 in long with or without a heel can be rooted in a frame in July. Ripened wood cuttings about 6 in long may be planted in a cold frame in November or in a sheltered spot in the open.

Eccremocarpus *E. scaber* (Chilan Glory Flower) is easily raised from seed, which is abundantly produced. Sow in pans and leave outside over winter. In February bring indoors for quick germination.

Elaeagnus Deciduous species can be raised from seed stratified about 18 months and sown in early spring in heat. Softwood cuttings of *E. multiflora* and *E. glabra* will root in a close case or frame. Firm wood cuttings 4 to 5 in long with a heel can

90

be struck in a cold frame. Layering is approved for several difficult species such as *E. angustifolia* (oleaster), *E. macrophylla* and *E. umbellata*.

Erica (Heath) The usual method is by cuttings of semi-mature tips about 1½ in long. Insert these in a compost of peat and sand in small pots and place in a close case. Firm cuttings can also be rooted similarly taken in October–November. Layering is another method and many species can be easily divided in spring. Seed is used for a few species such as *E. arborea* (Tree Heath) and *E. lusitanica* (Portuguese Heath). It should be sown on sifted peat in February.

Escallonia Softwood sideshoots 2 to 4 in long root in a close case and mature shoots of similar type can be inserted in a cold frame in October, or even outside in a sheltered position.

Eucalyptus (Gum Tree) Easily raised from spring-sown seed in moderate heat.

Euonymus Softwood cuttings of *E. japonicus* with or without a heel strike in a close case in summer. Ripened wood with a heel, planted October, can be rooted in a cold frame. *E. radicans* can be rooted from softwood cuttings. Division is suitable for some varieties and several species can be raised from seed sown in spring.

Fatsia (Fig leaf Palm) Usually raised from seed, but cuttings also feasible.

Forsythia *F. suspensa* (Golden Bell) from hardwood cuttings 6 to 8 in long and planted in a cold frame in October or in a sheltered site in the open. Semi-mature tips root readily in a close case in summer.

Fothergilla (American Witch Hazel) Increased by layering autumn or spring. Cuttings of semi-ripe wood will strike in late summer in moderate heat.

Fuchsia Cuttings made from soft tips when available can be rooted in a close case at any time of the year.

Garrya Layering in the autumn is the usual method. Firm sideshoots 3 to 4 in long with a heel will root when inserted in a cold frame in October.

Gaultheria Several species including *G. shallon* are raised from seed sown in February. Layering and division are other suitable methods. Certain species such as *G. hispida* (Snowberry) and *G. oppositifolia* can be raised from cuttings secured in August and planted in a frame.

Genista *G. hispanica* (Spanish Gorse) and other species are raised from cuttings of ripe sideshoots 2 to 4 in long planted in a cold frame in September. Several species may be raised from seed sown under glass in February.

Griselinia Readily increased by sideshoots with a heel inserted in a cold frame in October, or softwood cuttings in a frame or close case in summer.

Hamamelis In nurseries, grafting under glass in August is the usual method of increase. The stock used is *H. virginiana* (Witch Hazel) which is raised from seed sown under glass. Layering in spring is the most suitable method for the amateur.

Hedera (Ivy) Ivy is normally increased from cuttings, either the softwood type taken in summer and inserted in a close case or frame, or firm nodal cuttings planted in a cold frame in October.

Hibiscus Seed is sown in the open in spring. Varieties are root-grafted on *H. syriacus* in early January under glass.

Hibiscus (Rose Mallow) The shrubby species of *H. syriacus* are usually propagated from hardwood cuttings in early autumn with bottom heat. Layering and autumn-sown seed are other methods and grafting on seedling stocks in heat in January.

Hippophae (Sea Buckthorn) Increased by root cuttings outdoors in spring by autumn-sown seed and by layering.

Hydrangea *H. macrophylla*, (syn. *H. hortensis*), and its varieties are best propagated from softwood cuttings taken from plants grown indoors or outdoors. The cuttings are potted singly into small pots, using sandy compost. The pots are placed in a close case, *H. paniculata* and *H. arborescens* are usually mound-layered in the autumn. *H. petiolaris* is raised from seed sown in heat in spring.

Hypericum The species are raised from seed stored dry and sown in spring under glass. Cuttings are also generally used either of firm wood with a heel inserted in a cold frame in autumn or softwood cuttings in a close case in summer. *H. calycinum* (Rose of Sharon) is easily increased by division.

Ilex (Holly) The common holly and other species are raised

92

Cyclamen neapolitanum is normally increased by seed which is freely produced. Old corms can also be increased by cutting into several pieces each with an 'eye'.

from seed stratified for 18 months and sown outside in spring. Layering in the autumn is another method. For this purpose, stock plants must be partly lifted and laid on their sides, and each young shoot should be tongued before being layered. Budding of varieties on the type plant is also done.

Jasminum All species and varieties are raised from cuttings. Hardwood cuttings with a heel are taken in November and inserted in a cold frame, or in a sheltered position in the open. Immature sideshoots also strike readily in a close case or frame in summer.

Kalmia Seed is sown in peaty compost under glass in March. Layering is the best method for increasing varieties. It should be done in the autumn, the layered shoots being twisted or tongued. *K. polifolia* will grow from semi-mature cuttings taken in August and inserted under bell glasses.

Kerria (Jew's mallow) Cuttings of firm shoots with a heel root when planted in the open in autumn. Softwood tip cuttings will also strike in a close case in July.

Kolkwitzia amabilis Take semi-hardwood cuttings in July and insert in a close case.

Laburnum *L. anagyroides* (Common Laburnum, Golden Chain) is easily raised from seed sown in the open in spring. Hardwood cuttings 9 to 12 in long with a heel are used for all varieties. Varieties may also be budded or grafted on common laburnum.

Laurus *L. nobilis* (Bay Laurel) is normally raised from cuttings of firm shoots with a heel planted in a cold frame in early winter. Layering is another method sometimes used.

Lavatera (Tree Mallow) Shrubby species are increased by seed in gentle heat in spring. Summer cuttings will also strike.

Leycesteria Easily increased from summer cuttings in a shaded frame, also by spring-sown seed.

Ligustrum The ordinary privet *L. vulgare* is very easily propagated from hardwood cuttings which are made 10 to 12 in long and inserted in the open in autumn and winter. Various other species and varieties are best increased from softwood cuttings about 3 in long with a heel, and planted in a close case.

Lippia (Sweet scented verbena) Shrubby species are easily increased by summer cuttings.

94

Lonicera The climbing types are increased by ripe cuttings about 6 in long inserted in a cold frame or by softwood tips, 3 in long with or without a heel inserted in June in a close case. The shrubby species are increased similarly. *L. nitida*, the popular evergreen hedging plant, is readily increased by autumn cuttings planted in the open.

Magnolia Layering is probably the best method and should be done in spring. Only young shoots should be layered after being slit to form a tongue. Many species can be raised from seed which should be sown when ripe in the open. Small quantities may be sown in boxes or pans, left outside until February and then brought indoors. Varieties are grafted under glass on seedling stocks.

Mahonia Raised from seed in the same manner as berberis. The valuable species, *M. bealei*, may be propagated from cuttings. These are taken in June or July, made about 6 in long and inserted in a close case.

Malus (Apple) The flowering apples are usually grafted onto seedlings of *M. pumila* (Wild Crab); see also Pyrus.

Olearia The species and varieties are readily increased from softwood sideshoots in summer in a close case, or from mature laterals with a heel secured in October and planted in a cold frame.

Osmanthus Cuttings of mature sideshoots with a heel taken in October will strike in a cold frame. Most of the species are, however, layered in the autumn, the young shoots being well tongued before pegging down.

Osmarea Increased by summer cuttings.

Passiflora (Passion Flower) Increase is by seed or summer cuttings.

Pernettya Variable plants are produced from seed sown in the spring in the open. Varieties are raised from cuttings of small sideshoots taken in July or August and inserted in a close case. Division is an easy method for the amateur and should be done in spring. Young shoots may be twisted and layered in autumn.

Perowskia This is increased from semi-mature shoots without a heel secured in July or August and inserted in a close case.

Philadelphus All the species are easily increased from cuttings of ripe shoots about 9 in long with a heel planted in the open in a sheltered position is a well-drained light soil. Softwood sideshoots 3 to 4 in long taken in July will strike in a frame or close case.

Phlomis *P. fruticosa* (Jerusalem sage) and other species will root from cuttings of ripe shoots about 3 in long when planted in a cold frame in October. Softwood nodal cuttings also strike readily in a close case. All may be raised from seed.

Pieris Cuttings of semi-mature sideshoots about 3 in long with a heel may be rooted under bell glasses when inserted in August or September. Autumn layering is another method.

Piptanthus Sow seed in boxes or pans in heat in spring.

Pittosporum The species are raised from seed sown under glass in March in a peaty compost. Cuttings of semi-mature shoots with a slight heel will root in a close case in July. Varieties are grafted onto seedlings of the parent species under glass in winter.

Polygonum The popular climber *P. baldschuanicum* is propagated from hardwood cuttings 6 to 8 in long with a heel. These should be potted singly into 3 in pots which are stood on a glasshouse staging or cold frame.

Populus (Poplar) Hardwood cuttings about 8 in long, planted in a light sandy soil in the autumn, strike readily. Some species can be increased from suckers or root cuttings. Certain varieties must be grafted, and this is usually done in winter on the stock *P. canescens* (Grey Poplar).

Potentilla The shrubby species such as *P. fruticosa* may be raised either from small softwood sideshoots with a heel inserted in a close case in July or from hardwood sideshoots planted in a cold frame in October.

Prunus All these are budded or grafted onto various stocks in a similar manner to fruit trees. Budding is usually done in July or August and grafting in April in the open on established stocks. The stocks used include *P. Avium* (Sweet Cherry) which is preferred for the Japanese cherries; *P. cerasifera* (myrobalan) is used for the varieties of this species: seedling peach for the flowering peaches and Brompton or St. Julian for *P. salicina* (syn. *P. triloba*) and its varieties.

Pyracantha (Fire Thorn) The species and varieties are increased by inserting small semi-mature sideshoots with a

96

heel in a cold frame in August.

Pyrus (Pear) To produce pear seedlings, the pips should be stratified over winter and the seed sown in spring in the open or in boxes in a cold frame or glasshouse. Other species and varieties may be grafted onto these seedlings in March in the open.

Rhododendron This very large genus is increased by various methods. Seed is used to raise *R. ponticum* (common rhododendron), which is widely used as a stock for grafting. The seed is collected when ripe and sown in the open in spring or in pans or boxes in a cold frame. A very wide range of species are also increased by seed. This requires great care in sowing, which is done in pots or pans, containing sifted peat. In nurseries, a large number of varieties are grafted, this being done in winter under glass. Layering is also widely used in nurseries, stock plants being established in peaty soil. Young shoots are layered in autumn and before each is pegged down should be well twisted. Cuttings are used mainly for the smaller species such as *R. impeditum* and *R. racemosum*, and also for the evergreen azalea section. Semi-mature sideshoots are taken in June or July and inserted in pots or pans, using a compost of peat and sand. The receptacles are placed in a close case.

Rhus *R. typhina* (sumach) is increased by root cuttings about 1¹/₂ in long which are secured in winter and inserted singly in small pots. *R. cotinus* (smoke tree) comes true from seed which is sown in the open in autumn. Varieties of this species are mound-layered in autumn.

Ribes The flowering currants are easily propagated from hardwood cuttings about 6 in long inserted in the open in the autumn.

Romneya (California Tree Poppy) The species, *R. coulteri* and *R. trichocalyx* are propagated from root cuttings which are made into lengths of about 1 in and are inserted horizontally in pots or boxes of sandy compost in winter.

Rosmarinus (Rosemary) Readily increased from cuttings in August, inserted in a cold frame.

Rubus The ornamental brambles are increased by division or by tip-layering. *R. cockburnianus* (syn. *R. giraldianus*) is increased by root cuttings 1¹/₂ in long, which are inserted singly in pots placed under glass.

Ruscus (Butcher's Broom) Propagate by dividing the creeping root stock in spring.

Salix (Willow) One of the easiest plants to increase from almost any type of stem cutting.

Salvia (Sage) The shrubby species are easily increased from softwood cuttings in a close case.

Santolina chamaecyparissus (Lavender Cotton) Softwood sideshoots about 3 in long taken in July root readily in a close case.

Sarcococca Easily increased by division, autumn cuttings and seed in spring.

Schizophragma (Climbing Hydrangea) Usually increased from cuttings with a heel in July in a close frame.

Senecio The shrubby species such as *S. greyii* and *S. laxifolius* root from cuttings of semi-mature sideshoots in a close frame in August. Hardwood sideshoots can be rooted in a cold frame when planted in the autumn.

Skimmia Ripened sideshoots taken in the autumn strike in a cold frame but are rather slow. Layering in the autumn is another method.

Solanum The shrubby climbing species such as *S. jasminoides* are readily increased from cuttings in early summer.

Spartium *S. junceum* (Spanish Broom) is easily increased from seed sown in spring under glass.

Spiraea All the spiraeas can be propagated from cuttings either of the hardwood type taken in the autumn about 8 in long with or without a heel and planted in the open, or softwood cuttings struck in a close case or frame. Division is used for the dwarf stooling kinds such as *S. douglasii* and *S. japonica* and its varieties.

Symphoricarpus (Snowberry, St. Peter's Wort) Easily increased by hardwood cuttings taken in October and planted in the open or by division.

Syringa (Lilac) Tongued young shoots are layered in the spring. Varieties are also grafted on seedlings of *S. vulgaris* (Common Lilac). This is done under glass in the spring. Privet, which is sometimes used for grafting on, is a bad stock. Some varieties can be rooted from half-ripened cuttings inserted in a cold frame in July.

Tamarix Cuttings made from hardwood 8 to 10 in long are planted in the open in sheltered positions in autumn.

Teucrium The shrubby species root from small soft side-shoots taken in June or July and planted in a close case or under bell glasses.

Ulex (Gorse) Stratified seed sown outside in spring or in pots placed in a cold frame germinates freely. Sideshoots taken in July are inserted in a frame and kept syringed frequently.

Veronica Semi-mature sideshoots about 3 in long with a heel root readily in a cold frame in July.

Viburnum Several species are increased from seed sown in pots in spring and placed in a cold frame or glasshouse. Layering is another method used for *V. opulus* (Guelder Rose) and its varieties. Semi-mature shoots of certain species such as *V.X. burkwoodii*, *V. carlesii*, *V. fragrans* and *V. tinus* (Laurustinus) strike readily in a close case in July or August.

Vinca (Periwinkle) Division in early spring is a simple method. Soft tips root easily in a close case in summer and cuttings of ripe shoots can be struck in a cold frame in the autumn.

Vitis (Vine) Varieties of the grape vine and *V. coignetiae* can be increased from hardwood cuttings 6 to 8 in long, inserted in a cold frame in October.

Wisteria *W. sinensis* can be raised from seed sown in heat in early spring. Varieties are root-grafted onto these seedlings when established in winter. Young shoots may be layered into pots in the autumn.

Yucca Root cuttings 2 to 3 in long are secured in winter and inserted in boxes of sandy soil which are placed in heat.

17
Fruit

All the berried fruits (also known as soft fruits or bush fruits) are grown on their own roots and as a general rule are easy to propagate.

Blackcurrants *Ribes nigrum* The usual method of increase is by hardwood cuttings which should preferably be secured immediately after leaf-fall. It is important to take cuttings only from healthy, heavy-cropping bushes and to avoid those that show symptoms of 'big bud' or virus in the leaves.

Select firm, well-ripened shoots of the current season's growth, but if these are hard to find, 2- or 3-year-old wood may be used. Cut the shoots about 10 in long at a leaf joint. Select a well-drained site and insert the cuttings about 6 in deep in the soil. When growth commences the following spring, the topmost buds produce shoots which arise at or just below the soil surface, forming a 'stool' type of bush without a main stem or 'leg'.

In the autumn, the young plants may be moved to a nursery bed about 18 in apart or to permanent positions at 3 ft spacings. After transplanting, cut back the young growths to within 1 or 2 buds of their point of origin.

Blackcurrants can also be easily propagated by mound layering. The method is to cut a plant intended for propagation back to near ground level. The young shoots that arise from the base are kept mounded up as they grow and will produce roots below the soil. These rooted shoots may be pulled off and transplanted in the autumn.

100

Red and whitecurrants *Ribes sp.* Normally these are also increased from hardwood cuttings, firm 1-year-old stems being preferable. Take 10 to 12 long cuttings in the autumn and cut off all buds with a sharp knife except the top 3 or 4. This means that shoots will be produced only from near the tip of the cutting, to eventually form the head of the bush. Unlike blackcurrants, the red and white types are best grown with a clean stem or leg.

Gooseberries *Ribes grossularia* The method is the same as for red and whitecurrants, making a bush with a clean stem or leg.

Raspberries *Rubus idaeus* These may be said to propagate themselves with very little assistance from the gardener. Often, indeed, the number of new canes that arise naturally alongside the permanent row are an embarrassment. These young canes may be dug up and transplanted in the autumn or winter and afterwards cut down to within about 9 in of ground level. Raspberries may also be increased from root cuttings.

Blackberries and logan berries *Rubus sp* These and related fruits are normally increased from tip layers, and this may be done in summer. Light soil well supplied with organic matter is most suitable. Make a slit in the soil with a spade or trowel and insert the tips of young canes in this to a depth of about 6 in. The soil is then trodden firm. New plants can be severed from their 'parents' and transplanted the following spring.

Strawberries *Fragaria* Runners are the natural and speedy method of increase. After the fruit has been picked and the bed weeded, the runners may be pegged down to enable them to root more quickly. These may be transplanted from midsummer to autumn, but the earlier the better as runners planted in summer often produce a number of fruits the following season, whereas this is unlikely with later planting.

Tree fruits

The techniques of budding and grafting described earlier are the methods used for 'making' fruit trees, but they demand a fair degree of expertise and preparation and are generally left to a skilled nurseryman. A scion, which eventually becomes the tree head, is budded or grafted to a rootstock which supplies the root system. The rootstock has considerable influence on the rest of the tree, particularly during the early years, affecting rate of growth, height and spread and the age when the tree commences fruiting.

Fruit growers use grafting techniques to completely change the variety of an apple or pear tree, and nurserymen are able to supply trees on which different branches produce different varieties of the same fruit.

18
Raising vegetables and herbs

Most vegetable crops are produced from seed but a few of the perennial kinds, such as rhubarb and seakale, are propagated by vegetative means. Seed is sown in drills which facilitates weed control. The following alphabetical list shows how the different kinds are increased:

Artichoke, Globe *Cynara scolymus* Usually increased from suckers which are found at the base of established plants. These are simply cut off with a sharp knife and planted in the open 4 in deep and about 2 ft apart.

Artichoke, Jerusalem *Helianthus tuberosus* These are very hardy and increase naturally by tubers. Plant from February to April in rows 30 in apart, allowing 12 in between the tubers.

Asparagus *Asparagus officinalis* Asparagus is raised from seed, which may be sown in March or early April in drills 18 in apart. After covering with soil the bed should be rolled. The seed is slow to germinate. The seedlings are thinned out to 3 in apart leaving the strongest. During the summer, hoe frequently, and watering in dry weather is beneficial. The following spring the plants may be moved to their permanent site.

Basil. Sweet *Ocimum basilicum* This is an annual herb which is raised by sowing seed in heat in April and planting out in early June.

Beans, Broad *Vicia faba* For early crops, seed of the long-pod varieties are sown in November, except in cold districts. A sowing may also be made on a warm border in January or February. Another method is to start the plants in a cold frame and transplant them later. Alternatively, early sowings may be protected by covering the rows with cloches for a period. Beans are usually sown in double rows, the seed being spaced 9 in apart each way. The double rows should be 2 ft 6 in apart.

Beans, Kidney or French *Phaseolus vulgaris* Sow in the open in late April or early May in drills 2 in deep and 18 in apart. Space the seed 3 in apart and thin out to 6 in apart. Early crops are secured by sowing in March in cold frames in drills across the frame 1 ft apart. When the plants are through, cover the frames with mats on frosty nights. Double rows sown in April may be given protection for a period with cloches. Treat as tender in US.

Beans, Runner *Phaseolus multiflorus* Sow in the open in early May. Earlier sowing may be done in glasshouses, or frames for transplanting and sowings in the US should not be made until after the last frost. For training on sticks or poles the seeds are spaced $4^1/_2$ in apart in double rows 9 in apart the double rows being not less than 6 ft apart. The plants are usually thinned out to leave 9 in between them.

Beetroot *Beta vulgaris* Sow in late April or early May after the last frost in the US in rows 12 in apart. Thin out the plants to about 2 to 3 in apart. Wider spacing usually results in poorer quality and lower yield.

Borage *Borago officinalis* This annual herb may be sown in March or April, and thinned out to 12 to 15 in apart.

Borecole or Kale *Brassica oleracea acephala* Sow thinly in early April in rows 12 in apart. Transplant when large enough, leaving 2 ft 6 in between the rows and spacing the plants 2 ft apart.

Broccoli *Brassica oleracea botrytis asparagoides* Sow according to variety from mid-April to mid-May in drills as advised for borecole. Plant out from June to mid-July. Space the plants about 2 ft 3 in apart each way.

Brussels Sprouts *Brassica oleracea bullata gemmifera* For early crops, the seeds may be sown in a glasshouse or cold frame, the plants being set out in April or May. In some

104

districts, autumn sowings are made in sheltered borders for spring transplanting; seed for the main crop is usually sown in drills in March. These are transplanted in June. Set the plants 3 ft apart each way.

Cabbage *Brassica oleracea capitata* Summer cabbage of the round Primo type is usually sown in a cold frame in January.

Carrots *Daucus carota* For early crops, the seed may be sown broadcast thinly in a cold frame in October or January and left there until ready for pulling. The stump-rooted type may be sown in a warm border in March. Main crop carrots are usually sown in April or early May. Make the drills $1/2$ to 1 in deep and 12 to 18 in apart. When sown evenly and not too thickly there is no need to thin out.

Cauliflower *Brassica oleracea botrytis cauliflora* Early cauliflowers are usually sown in September in a cold frame. When large enough to handle, they are pricked-off in another frame and wintered there. Alternatively these plants may be potted into 3 in pots which are placed either in a cold frame or on a glasshouse floor and wintered there. These plants are set out

Seedling cauliflowers. Note the greater development of some than others. It is the weaker ones that should be removed when thinning.

105

pricked-off into other boxes, hardened off and planted in the open in April. Further sowings for autumn produce may be made outside in April.

Celery *Apium graveolens* Sow in a heated glasshouse in February or March and prick-off into boxes or a cold frame. Harden-off and plant out in trenches (or in beds for self-blanching celery) in late May or early June after the last frost in the US. Seed may also be sown in a cold frame in March. Pelleted seed can be used.

Celeriac or Turnip-rooted celery *Apium graveolens rapaceum* Sow about mid-March and give the same treatment as advised for celery. Plant out on the flat 12 in square in early June.

Chicory *Cichorium intybus* Sow in late May in rows 1 ft apart. Thin the plants out 6 to 9 in apart.

Chives *Allium schoenoprasum* Lift and divide the clumps every 3 or 4 years in the spring. May also be grown from seed.

Endive *Cichorium endivia* This autumn and winter salad crop should be sown for succession from June until mid-August. Spacing is similar to lettuce.

Horseradish *Cochlearia armoracia* Easily propagated from root cuttings about 3 in long. These are inserted in holes made with a long dibber and afterwards filled in with soil. Plant in spring 18 by 12 in apart.

Leeks *allium porrum* The seed is usually sown about mid-March in shallow drills 15 in apart. The seedlings are transplanted when large enough to handle 6 in apart in deep drills made about 12 in apart. To secure plants earlier a sowing may be made in a cold frame in January or February.

Lettuce *Lactuca sativa* With the aid of glass protection and artificial heat it is possible to have lettuce for cutting all the year round. Suitable varieties must be used for the different seasons and conditions. Approximate dates of sowing the principal varieties are as follows:

1. Sow outdoors from early March until late July at 2 to 3 week intervals, using varieties of the Trocodera or Webb's Wonderful or Avon Defiance.

2. Sow the varieties Winter Crop, Imperial or Arctic King about mid-August in the open. Transplant in late September in a sheltered situation.

106

3. Sow the varieties Seaqueen or Emerald in a cold frame or glasshouse in early September and a further sowing of the same varieties in October in slight heat. Transplant the seedlings, when large enough to handle, in a heated glasshouse border.

4. Sow the variety May Princess in a cold frame in early October. Transplant the seedlings in December or January in a cold frame or glasshouse. Pelleted seed allows spacing.

Lettuce are normally grown in rows from 9 to 12 in apart, with the same distance left between the plants.

Marjoram, Pot *Origanum onites* A perennial herb which is increased by sowing seed in April. Plant about 9 in apart.

Marjoram, Sweet *Origanum majorana* This herb is treated as an annual and is usually sown in heat in March or April. Transplant in May 9 in apart.

Mint *Mentha sp.* Easily propagated by division of the underground stems in spring or autumn. Summer cuttings also root in a shady situation in the open.

Onions *Allium cepa* Bulb onions are produced by sowing in the open in March. Earlier sowing may be made in heat and the plants set out when large enough to handle. A sowing can also be done in August and the seedlings transplanted the following spring. The variety White Lisbon is often sown in August for use as a salad in spring. Bulb onions are grown in rows 12 to 15 in apart, the plants being spaced about 4 in apart. Pelleted seed is available for space-sowing. Potato onions and shallots *Allium ascalonicum* are increased by division of the bulbs which are lifted in the autumn and planted in spring 6 to 9 in apart.

Parsley *Carum petroselinum* Sow in March for summer cutting, in June for winter produce, and again in August for use the following spring. The seed is slow to germinate. Drills should be spaced 12 in apart and the plants thinned to 3 to 4 in apart.

Parsnips *Pastinaca sativa* Sow in February or March in drills 15 to 18 in apart. Thin out plants to 6 in apart.

Peas *Pisum sativum* In mild districts, early peas may be sown in November. Normally successional sowings are made from early March until mid-June. Early sowings may be protected by cloches. Use early mid-season and later varieties according to season, but early varieties should be chosen for late sowing. Ordinary V-shaped drills about 2$\frac{1}{2}$ to 3 in deep, and spaced about the same distance apart as the plants are expected to grow, are suitable.

Radishes *Raphanus sativus* Successional sowings can be made from February to August in the open. Earlier sowings may be protected in cold frames or under cloches. Seed is usually sown broadcast or in drills 6 in apart.

Rhubarb *Rheum hybridum* Divide the old roots in autumn or spring. Each portion should have at least one crown or 'eye' and is called a set. The sets are usually planted in rows 3 ft apart with 2$\frac{1}{2}$ ft being allowed between the sets. Rhubarb can also be raised from seed, but does not come true to variety. Sow in March in drills 1 in deep and 12 in apart. Thin the plants to 9 in apart, and transplant in the autumn or spring.

Sage *Salvia officinalis* Easily increased from seed sown outdoors in April. Soft cuttings root readily in spring or early summer and plants so raised are often preferable to those from seed. Plant 15 in apart.

Salsify *Tragopogon porrifolius* Sow in drills 1 in deep and 12 in apart in April.

Scorzonera *Scorzonera hispanica* Treat in the same manner as salsify.

Seakale *Crambe maritima* Usually increased from root cuttings or thongs about $\frac{1}{4}$ in thick and 6 in long. These are prepared in the autumn, tied in bundles and laid in sand. The sprouts which appear on each cutting in spring should be rubbed off, except the strongest. The thongs are then planted about 1 in below the surface in rows 18 in apart and 15 in between the cuttings. Seakale may also be raised from seed sown in the open in March.

Spinach *Spinacia oleracea* A succession of summer spinach is maintained by sowing at two to three week intervals from early March until the end of June. Winter spinach is provided

Parsley comes readily and thickly from seed and needs thinning if the plants are to grow on well.

by making a sowing at about mid-August and another at mid-September. Sow in drills 1 in deep and 12 in apart, and thin to 6 in.

Spinach, New Zealand *Tetragonia expansa* Seed is sown under glass in late March and the seedlings are potted. Harden-off, and plant out in late May, the plants being 2 ft apart and the rows 3 ft.

Spinach Beet *Beta cicla* Two sowings are made for succession, one in april and one in late July. Sow in drills 15 in apart and thin out the plants to 6 to 9 in apart.

Thyme *Thymus vulgaris* The common thyme is easily raised from seed sown in April or by division in spring. Plant about 4 in apart.

Turnips *Brassica rapa* Turnips and swedes are raised by sowing seed from March until May. Earlier sowings may be made in frames or protected by cloches. The plants are usually thinned out to about 4 in apart.

Vegetable Marrows Squashes *Cucurbita pepo ovifera* Sow seed singly in 3 in pots in late March or early April. Germinate, and grow in heat for a period. Harden-off, and plant out towards the end of May. The seed may also be sown in a cold frame in April or in the open in May.

Half-hardy/tender vegetables

Tomatoes *Lycopersicum esculentum* This is one of the most popular and valuable crops for the garden and greenhouse. Outdoor varieties cannot safely be planted until risk of frost is past, and all should be raised from seed in a heated greenhouse enjoying good light. Use a reliable seed sowing compost and space seeds about $\frac{1}{2}$ in apart in trays or pots covered with about $\frac{1}{8}$ in of the same compost. During the night, maintain a temperature of 65°F, which may rise to over 70°F during the day. When the seed leaves have expanded, usually eight to ten days after sowing set the seedlings singly in $4\frac{1}{4}$ in plastic pots, using the potting mix of your choice. From that time the night temperature is lowered to 60°F. Water the young plants frequently in sunny weather, less frequently when dull.

When the seedlings are well established in their pots, they need to be fed with a liquid fertilizer formulated for tomatoes. Space out the plants as they grow so that the leaves do not overlap.

Planting out in the greenhouse border or into larger pots is best done when about half the plants have their first flower truss showing. Outdoor varieties should be accustomed to lower temperatures once established in their pots and can be hardened-off in a cold frame during April.

Cucumbers *Cucumis sativus* Propagation of cucumbers is somewhat less exacting than tomatoes. The best method is to sow the seed singly in 3 in pots, using a potting compost rather than a sowing compost in this case, as the cucumber is a gross feeder. Fill each pot and press a seed into the centre $1/2$ in deep. Give a good watering, then place glass and paper over the pots.

Maintain a temperature of about 70°F and a moist atmosphere, and germination occurs in 36 to 48 hours. Then remove the glass and paper but avoid exposing the seedlings to strong sunlight. Continue to grow in a temperature of 65 to 70°F and water when necessary.

When the roots have reached the sides of the pots, pot on into 6 in pots using a similar or richer compost, and once established, apply liquid feeds with some waterings. Support the plants as they grow with 2 ft canes inserted in the pots. Plant out in the greenhouse beds before there is any risk of a check to growth in the pots.

Cucumbers for outdoor culture–ridge cucumbers–do not possess the fine quality of the greenhouse type, but are raised from seed in the same way, given gradually cooler conditions, hardened-off and planted in rich soil when the risk of frost is past.

Melons *Cucumis melo* Melons may be raised in the same manner as cucumbers except that the mixes used need not be so rich. Apart from greenhouse cultivation melons are suited to being grown in frames or under cloches if planted out after spring frost.

Tomatoes are probably the most colourful and satisfying of all vegetable crops to grow. There are numerous varieties and all can easily be raised from seed.

19
How decorative glasshouse and room plants are increased

Under this heading a wide range of ornamental foliage and flowering plants is included. Most of them are increased by seed or cuttings involving the general principles outlined in previous chapters in relation to these methods. Thus seeds are usually sown in sowing compost and potted into potting compost with extra feeding according to the type of plant and the length of time it is likely to remain undisturbed, after potting.

The seeds of glasshouse plants are usually germinated in a temperature of 55 to 65°F, and a similar temperature is allowed for the propagation of cuttings.

Abutilon (Indian mallow) Hardwood cuttings in spring or autumn will root with bottom heat.

Acacia (Wattle) Insert half-ripe cuttings with a heel in July in pots containing a peaty compost. Plunge the pots in a close case; when rooted, pot off singly and grow in cool conditions. Seeds sown when ripe in a mixture of peat and sand germinate freely in a temperature of 60°F.

Acalypha (Copper leaf) Cuttings should be struck in brisk heat in spring.

Achimenes Sow seed in early spring with care, as it is very small. Germinate in a temperature of 60 to 65°F. Prick-off the seedlings in light peaty compost and grow in a similar temperature. Cuttings secured from plants started in heat should be inserted in a close case, using a compost of peat and sand with bottom heat. Leaf cuttings also root under similar conditions when their stalks are inserted in the compost. Another method is to rub the scales off the corms and sow them in the same compost, covering lightly. Given bottom heat the scales soon commence to grow.

Aechmea Readily increased from suckers which develop naturally. These should be potted and kept in a close warm atmosphere until rooted.

Agapanthus (African lily) Divide the plants just as growth begins in the spring, and pot the rooted portions.

Agave (American aloe) Offsets which are freely produced are a simple means of increase.

Anthurium Divide the roots in March when re-potting. Sow seed in a mixture of chopped sphagnum moss, charcoal and sand, temperature 70 to 80°F.

Aphelandra Increase from nodal cuttings of semi-mature wood or soft shoots with a heel in bottom heat of 70°F.

Aralia Soft cuttings made from sideshoots can be rooted in a close case. Root cuttings taken in spring will also grow. These are made about 2 in long and are inserted in pots of sandy soil. The pots are placed in a close case and bottom heat promotes rooting. Varieties and species can be grafted under glass on the stock *A. reticulata*.

Araucaria excelsa (Norfolk Island pine) Seed in heat is the usual method.

Ardisia (Spear flower) Seed may be extracted from ripe berries and sown immediately in heat. Cuttings from March to September will strike in heat.

Asparagus The well-known asparagus ferus *A. plumosus* and *A. sprengeri* are raised from seed. Sow in spring in sowing compost. Pot singly and grow in a fairly moist warm atmosphere. Roots may also be divided when re-potting large plants in spring.

Aspidistra Divide and re-pot during the growing season. Shade for a few days afterwards.

Auricula *Primula sp*. Named varieties are increased from rooted offsets secured when re-potting after flowering. These are inserted in small pots. Unrooted offsets will strike when inserted in a close case.

114

Begonia Most of the begonias may be increased from seed, which is sown in early spring. The seed is very small and requires a fine surface and little or no covering. Germinate in a temperature of 60 to 65°F. The ornamental-leaved kinds such as *B. imperialis* and *B. rex* are increased by leaf cuttings. The fibrous-rooted kinds, like *B. socotrana* and its varieties such as John Heal and Gloire de Lorraine, are readily increased from stem cuttings secured from plants started in heat in early spring. The propagation of tuberous-rooted begonias is mainly by division of the tubers.

Beloperone Easily increased from cuttings in moderate bottom heat.

Bignonia (Cross vine) These climbing plants are increased by soft cuttings in spring inserted in a close case. Young shoots may also be layered in the late summer.

Billbergia Increased by division the suckers being twisted off the main stem. Pat separately and keep in brisk heat.

Bougainvillea Cuttings taken with a heel in March or April strike in a warm case.

Bouvardia Increase is by soft cuttings about 2 in long taken in March and inserted in pots of sandy compost. Plunge the pots in bottom heat in a case. Root cuttings about 1 in long will grow when planted in pans in spring. Divide the plants when repotting.

Browallia Raised from seed sown in spring or in July for late winter and spring flowering.

Brunfelsia Cuttings root readily in moderate heat. Seed should be germinated in a temperature of at least 70°F.

Cacti and succulent plants This large and variable group of interesting plants is not difficult to propagate. Seed and cuttings are the principal methods. Seeds are usually obtained from the plant's native country, but is sometimes home-saved. Sowing may be done at any time if heat is available, but spring sowing is generally preferable. Sow thinly in well-drained pans of fine compost and cover according to the seed size. Plunge the pans in peat or ashes and cover them with paper to prevent rapid drying. Maintain a temperature of 50 to 55°F and remove the paper immediately germination is observed. Prick-off the seedlings in light compost as soon as they are large enough to handle.

Generally speaking cacti and succulents are very easy to root from cuttings but some species such as the spherical kinds, do not provide any material for this purpose. Any branching plant however will usually provide and offsets can be secured from others. Usually the cuttings should be dried for a few days before insertion. A compost of peat and sand is quite suitable but pure sand also serves the purpose. There is no need to provide close conditions but slight shade may be beneficial. Cuttings of strong-growing species such as the opuntias may be inserted singly in pots. Many kinds can be raised from leaf cuttings. Grafting is sometimes used for a few special kinds.

Caladium Propagated by division of the tubers in early spring.

Calathea (Zebra plant) Easily increased by division.

Cassia (Senna plant) Cuttings of half-ripe wood will strike. Seed is also used.

Calceolaria (Slipper flower) The shrubby species are increased by cuttings which are taken in autumn and inserted in a cold frame or in boxes kept in a cool glasshouse. Herbaceous calceolaria are raised from seed sown in July. Prick-off and pot on as required.

Callistemon (Bottle brush tree) Take cuttings of firm young shoots in summer and insert in pots containing a compost of peat and sand.

Campanula isophylla This species which is popular for hanging baskets should be raised annually from cuttings taken in summer.

Canna (Indian shot) Sow seed singly in small pots 1 to 2 in deep after soaking in water for 24 hours. Sowing is done in February or March, the pots being kept in a temperature of 70°F. Named varieties are raised by dividing the roots at potting time.

Carnation Perpetual flowering carnations are propagated by cuttings which should be selected with care. The best are those made from sideshoots found about the centre of the flowering stems. Cut below a node, remove the lower leaves and trim off the leaf tips. Insert the cuttings in sharp clean sand and maintain a temperature of 50 to 55°F. Bottom heat is also beneficial. Water after insertion, and subsequently keep the medium nicely moist. Keep the cuttings close at

116

first, but free ventilation should be given when rooting commences in a few weeks' time. Rooted cuttings are potted into 3 in pots using a gritty compost. Carnation cuttings of this type are normally taken from December to March, but they may be struck at other seasons.

Malmaison carnations are best increased by layering in June. The rooted layers are potted in August.

Celosia (Cockscomb) This annual makes an attractive pot plant. For this purpose the seed is sown in a warm house in March and is first pricked-off and then potted on as necessary.

Chlorophytum Readily increased by division.

Cineraria Raised from seed, which is sown for succession from April until June. Soilless mix is suitable for seed sowing. Pot the seedlings into mix and later into the same compost in larger pots.

Cissus Propagated by softwood cuttings about 2 in long with a heel. Bottom heat is an advantage.

Clivia Sow seeds in a warm house in spring and keep the seedlings growing until they flower. Division of old plants can be done when re-potting in February.

Cobaea scandens (Cup and saucer plant) Easily raised from seed sown in a warm house in early spring. The seedlings flower the same year. The variegated type is raised from cuttings of firm sideshoots taken in July or August and inserted in a close case.

Codiaeum (Croton or South Sea laurel) Easily increased by stem cuttings, which are inserted singly into pots. The latter are plunged in peat in a warm frame with brisk bottom heat.

Coleus (Flame nettle) Plants of variable types may be raised from seed sown in gentle heat in early spring. Cuttings of young shoots can be rooted at any period in sandy compost and kept close. Cuttings are essential to increase special forms.

Columnea Cuttings from firm shoots will strike in a close warm frame.

Cordyline The store and glasshouse species are increased by cutting the main stem into pieces 1 to 2 in long. These are inserted in a close warm case in sandy soil. The half-hardy species such as *C. australis* are easily raised from seed.

Crassula Sow seed in a warm house in early spring and keep the seedlings rather dry. Stem or leaf cuttings strike readily in pots of sandy compost stood on a glasshouse staging.

Crossandra Easily increased by cuttings with good bottom heat throughout the year.

Cryptanthus Offsets should be separated and potted up singly in the spring in brisk heat.

Cuphea (Mexican cigar flower) Increased from seed sown in March in gentle heat. Cuttings of young shoots can be struck in spring or summer.

Cyclamen Propagated by sowing seed in August or January. Sow each seed $1/4$ in deep and 1 in apart in pans. Allow a temperature of 50 to 55°F. Shade the seedlings from bright sun and pot when large enough to handle.

Cyperus (Umbrella plant) Propagated by seed or division in moderate heat.

Cytisus (Florists' genista) Usually increased by cuttings of sideshoots taken in spring with a slight heel and inserted in pots, which are placed in a close warm frame. Seed may also be used and is sown in spring.

Dieffenbachia Increased by basal suckers or by short stem pieces like cordylines in brisk heat.

Dracaena (Dragon plant) Propagated by stem pieces as for cordylines.

Erica (Heath) Tip cuttings 1 in long are inserted in pots in sandy peat in spring. Keep close and use gentle heat.

Euphorbia (Poinsettia) Cuttings are taken in early spring and struck in a close case with brisk heat.

Fatsia *F. japonica* can be raised from stem cuttings about 2 in long in spring. Insert in a close frame. Seeds sown singly in pots germinate at 65°F.

Ferns Division when re-potting, usually in spring, is generally adopted. The well-known *Asplenium bulbiferum* is increased by minute plantlets which develop on mature fronds. The fronds may be pegged down on light compost like a leaf cutting, or the plants may be taken off singly and carefully inserted round the sides of a pot.

Ficus (Rubber plant) Propagated by short nodal stem pieces, each with a leaf which are inserted in pots or boxes stood in a moderately heated glasshouse.

Freesia Separate the offsets when re-potting in the autumn. Usually 6 to 8 corms are inserted in a 5 in pot. Easily raised from seed sown when ripe or in April. Seedlings usually flower before they are one year old.

Fuchsia Usually raised from cuttings of soft young shoots. These are taken in spring or in autumn and inserted in pots of sandy compost. Seed may be sown in spring in pans and germinated in a temperature of about 60°F.

Gardenia These beautiful evergreen shrubs are increased by cuttings. Secure young shoots in January and insert singly in pots of sandy compost. Place the pots in a close case. Plants so raised will flower the following winter, but later-struck cuttings provide a succession of bloom.

Gesneria Can be increased naturally by division of the tubers. Cuttings made from young basal shoots will strike in early spring using brisk bottom heat. Seed is another method but requires care as it is extremely fine. Treat like begonias.

Gloxinia (Sinningia) The gloxinia may be increased by seed which is sown in pans of light compost in March. Basal cuttings are easily rooted in pots of sandy compost placed in a warm frame. Leaf cuttings provide another method.

Grevillea *G. robusta* is increased from seed sown in February in sandy compost in a temperature of about 70°F. Insert cuttings of young shoots taken with a heel in spring in pots of sandy compost. Place the pots in a close case.

Heliotrope Cuttings of soft young shoots are inserted in spring in well-drained pots of light compost. Place in a close frame until rooted and then pot singly. Pinch the young plants two or three times to induce a bushy habit. Cuttings to be trained as standards are not stopped until they reach the required height.

Hibiscus (Rose mallow) Cuttings from mature current season's growth will strike in bottom heat in the autumn. Other methods are autumn or spring layering, and grafting onto seedling stocks.

Hippeastrum Secure offsets when re-potting old bulbs in early spring. Seed should be sown in heat in March and potted on when necessary. Seedlings do not reach flowering age until they are three years old.

Hoya carnosa (Wax flower) Cuttings made from shoots of the

119

previous year's growth are inserted in pots containing a compost of peat and sand in the spring. The pots are placed in a frame kept at a temperature of 70 to 75°F. Young shoots may be layered during the summer by pegging them into pots of peat and sand.

Humea Sow seeds in pots or pans in July and place in a cold frame or glasshouse. Pot when large enough to handle.

Hydrangea *H. macrophylla, var. hortensia, syn. H. hortensia,* the common hydrangea, is increased by cuttings. Young shoots are taken in spring and inserted singly in small pots of sandy compost. Place in a warm frame until rooted, and afterwards, grow in a cool glasshouse or frame. Pot as required.

Impatiens (Balsam) Glasshouse species are easily raised from seed but special strains or varieties are readily propagated from softwood cuttings.

Ipomoea (Morning glory) Annual species are raised from seed sown in March in a warm house. Perennial species are increased from cuttings made from sideshoots during the summer. Strike in a close case. *I. batatas* (the sweet potato) is increased by division of the tubers in February or from cuttings.

Jasminum (Jasmine) Glasshouse species are propagated from soft young cuttings with a heel in spring. These are inserted in a warm frame.

Kalanchoe Seed may be raised in pots or pans of light compost in a temperature of 60 to 65°F. The seedlings require care in handling. Cuttings made from young side-shoots after flowering and leaf cuttings strike readily in sandy soil.

Lachenalia (Cape cowslip) Bulbs increase naturally and are usually re-potted in August and planted 1 to 2 in apart. Seeds may also be sown in the spring in heat and are not difficult to germinate. Leaf cuttings is another method.

Lantana Propagated by seed sown in spring in a temperature of 70 to 75°F. Cuttings of young sideshoots inserted round the sides of a pot in spring strike in a warm case. Half-ripened cuttings taken in August or September and treated similarly will also root.

Lapageria Can be raised from seed in pans and kept in a

120

warm house. Young shoots can also be layered indoors in spring or summer.

Monstera (Shingle plant) Cut the stem into pieces three joints long which will root in a warm frame.

Nerine (Guernsey lily) Offsets are secured when re-potting, usually in August or September, and are potted in a 3 to 6 in pot.

Nerium (Oleander) Cuttings made from firm wood are potted singly in sandy compost in spring or summer. Keep in a close frame until rooted.

Orchids These comprise a large and varied group of plants, and propagation must be related to the usual cultural practice for each kind. Many orchids can be increased by division, this being done when re-potting, which is normally undertaken just when growth is commencing in the spring. Genera treated in this way include Cattleyas, Calanthes, Cymbidiums, Cypripediums, Dendrobiums and Masdevallias. Propagation by seed is very difficult and undertaken by specialists only.

Palms These plants are usually increased by seed which is sown in spring or when available. Sow in well-drained pans or pots, which should be plunged in peat and kept in a propagating frame with a temperature of 70 to 75 °F. A few species can be propagated from suckers, which are potted in spring.

Passiflora (Passion flower) Cuttings 4 to 6 in long with a heel are made from young shoots in spring. These are inserted in a warm propagating frame. Seed is another method.

Pelargonium (Geranium) The different types are raised from cuttings which are cut off below a joint and are made 4 to 6 in long. Pot singly in small pots using sandy compost and place on a glasshouse staging. When rooted, the young plants are potted on as required. Some of the species may be raised from seed.

Peperomia Cuttings of short stem pieces with a single joint will root in pots stood on a glasshouse staging. Under closer more humid conditions damping-off is probable. Seed is another method.

Philodendron Increase by cuttings as for *Monstera*.

Plumbago (Cape lead-wort) Easily propagated by stem or

root cuttings with bottom heat.

Primula Seed is the normal method of increasing glasshouse primulas and is usually sown in April or May for winter flowering. *P. malacoides,* however, is sometimes sown in July to provide flowers in spring. Sow the seeds in pots and prick-off the seedlings when they are large enough to handle. Later, they are potted into 3 in pots and finally into the 5 in size. A temperature of 55 to 60°F is suitable for seed germination and 50 to 55°F for growing the seedlings. Double-flowered primulas are raised from cuttings or division.

Saintpaulia (African violet) Raised from seed which is usually sown in early spring. Sow in pots and keep them in a temperature of 65 to 70°F. Prick-off and pot as the seedlings grow. Shading is important. Leaf cuttings provide another method of increase.

Salvia *S. splendens* is usually raised from seed which is sown in the spring in pots or boxes. Germinate in a temperature of 60 to 65°F and pot singly in 3 in pots when the seedlings are large enough to handle. Most salvias may be increased from cuttings. These are secured from plants of the previous year retained over winter in mild heat and started into growth in spring. Insert the cuttings in pots which are placed in a propagating frame.

Sansevieria Easily increased by suckers. Also by leaf cuttings, the long leaves being cut into lengths of about 3 in, which are inserted in sandy compost in heat.

Saxifraga sarmentosa (Mother of thousands) This species increases naturally by creeping stems or runners. From these, the plantlets are removed and inserted in small pots.

Schizanthus (Butterfly flower) For summer flowering, sow the seeds in early spring in a warm house, August is the time to sow when the plants are required to bloom in the spring. Prick the seedlings off and pot as necessary.

Selaginella (Creeping moss) Cuttings inserted in pots strike readily in a close frame. When rooted, several cuttings may be potted together to form a single specimen. Division when re-potting is another method of increase.

Solanum *S. capsicastrum,* the popular winter cherry, is usually increased from seed which is sown in February in heat using seeding mix. Prick-off the seedlings and pot when they are large enough, first into 3 in pots and finally into the 5 in size. Pinch the plants to induce a bushy habit. The winter cherry may also be increased by cuttings inserted in pots placed in a propagating frame in March. The climbing species such as *S. jasminoides* are increased from soft young cuttings in spring struck in pots in a close case.

Strelitzia (Bird of paradise flower) May be raised from seed which, when sown, should be kept in a high temperature, preferably with bottom heat. Old plants may be divided when being re-potted in spring.

Streptocarpus (Cape primrose) Normally raised from seed which is sown in pans in January or February for winter flowering. Seed sown in July will flower the following summer. Germinate in a temperature of 60 to 65°F and prick-off and pot as necessary. Special forms of Streptocarpus may be increased by dividing the old plants. Leaf cuttings will grow if inserted in a sandy compost.

Streptosolen *S. jamesonii* is increased by cuttings made from soft young shoots in spring or summer and inserted in a close case. Rooted cuttings are potted singly and pinched once or twice to encourage a bushy habit.

Tradescantia (Spider-wort) Easily increased by cuttings of young shoots inserted in a close frame.

Thunbergia Propagate by seed sown in heat or by semi-hardwood cuttings in a warm frame.

Torenia Usually raised from seed sown in March or April and treated as a glasshouse annual. It may, however, be increased by cuttings in a close case.

Zantedeschia (Arum lily) Re-potting and division are usually done in August or September. Suckers which are readily produced are potted singly and started in mild heat. Seed provides another method and may be sown in spring in a warm glasshouse.

Zebrina Easily increased from cuttings in moderately warm conditions at any time.

Index

124

126